Pruning and training
fruit trees

PRACTICAL HORTICULTURE

Pruning and training fruit trees

Warren Somerville

INKATA PRESS

INKATA PRESS

A division of Butterworth-Heinemann Australia

Australia
Butterworth-Heinemann, 18 Salmon Street, Port Melbourne, 3207

Singapore
Butterworth-Heinemann Asia

United Kingdom
Butterworth-Heinemann Ltd, Oxford

USA
Butterworth-Heinemann, Newton

National Library of Australia Cataloguing-in-Publication entry

Somerville, Warren,
Pruning and training fruit trees.

Includes index.
ISBN 0 7506 8931 5.

1. Espaliers. 2. Fruit - Pruning. I. Title. (Series :
Practical horticulture).

634.044

Typeset by Ian MacArthur, Hornsby Heights, NSW.

Printed in Australia by Ligare Pty Ltd, Riverwood, NSW .

Contents

Introduction

Pruning and tree training are only two pieces in the large and complex jigsaw of commercial tree fruit production. But these are the two areas of orchard management which are commonly least understood. Most

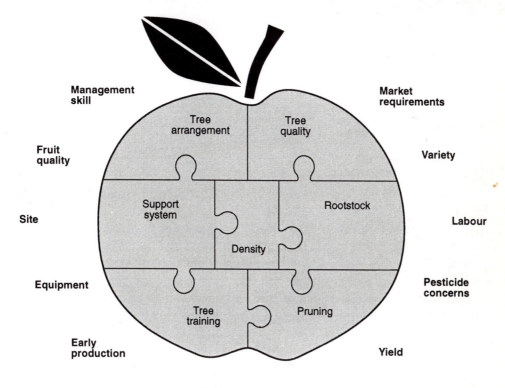

The orchard system puzzle

*Apple harvest in Washington State, USA. Red Delicious on
semi-dwarfing rootstock and trained to central leader*

successful growers are competent pruners and have achieved this skill by
observing over many years the results of the previous year's pruning and
training. The more years of experience with the same type of tree on the
same type of rootstock, managed with the same training system, the better
their skill.

When a grower accepts advice that he should be growing trees at very
much higher density using a different training system and perhaps with
training aids or tree support, he feels less confident of his skill and under-
pinning knowledge.

He may read reports that the HYTEC apple training system in Washington State is the latest or the Spanish bush system of training cherry trees is
the way to go. He may be wondering whether palmette is the best training
system for peaches or whether the open centre vase produces better fruit.
What really is the difference in tree management between spur type Red
Delicious and lateral bearers? And are all spur types the same?

He may read in the *Good Fruit Grower* that the Dutch five row bed system is amazing for its early production and high quality fruit.

Or he may be concerned that he cannot get the consistent good fruit
colour that peaches need for top market returns. Will a change in pruning
strategy help? And can a different pruning or training system succeed in
increasing cherry size, when heavy fertilising has not?

This book is based on the philosophy that the orchardist must go beyond field experience and understand how a tree grows and how it reacts to pruning in order to be a good pruner. So the discussion of how trees grow and the concept of apical dominance is the pivotal part of this book.

Deciduous fruit trees, their varieties — especially high chill varieties — and rootstocks are used as examples to illustrate points. This reflects the major experience of the author. There is also an immense amount of research in deciduous fruit trees from around the world to draw from. However, the concepts discussed apply to all tree fruit crops as well as to vine and cane crops.

Quality cherries at Young, NSW.

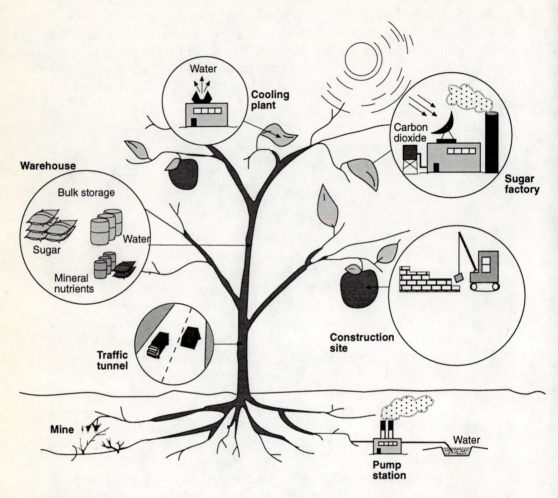

Figure 1.1 *A fruit tree can be likened to a community in which the primary industry is the production of fruit*

CHAPTER 1

The fruit tree

A fruit tree can in many respects be likened to to a self sufficient community where each part plays an important role and depends on every other.

The major roles of the roots are to "mine" nutrients and to extract water from the soil. These two resources are then transported by the traffic system to those parts of the plant which use them.

The leaves are the major users of water and nutrients. Much of the water is evaporated from the leaves in the process called **transpiration**. Transpiration is the driving force of the pumping system which distributes materials throughout the plant. Transpiration also assists in keeping the plant cool.

The most important function of the leaves however is **photosynthesis**. The leaves intercept sunlight and absorb carbon dioxide. With these and the water and nutrients the leaves then manufacture sugars. The sugars are the basic building blocks from which all other plant materials are manufactured.

Most of the sugars and other materials are used up by the growing processes of the tree. Much of this is used in the production of new materials such as shoots and fruit. All living cells use some sugar in the process called **respiration**. This process provides the energy to keep life going and for all of the chemical conversions which occur in the cells. Some of the sugar is also used to provide the energy to the cells in the roots to support their mining and water pumping functions.

Some of the sugar, nutrients and other materials are stored in the stems and in the roots. The plants survive the winter, commence growth, flower and begin the development of the fruit in the spring at the expense of this stored material.

For commercial fruit production clearly we must maximise all of the above operations. Pruning and tree training are part of the management strategies used to achieve this.

The stock and the scion

Most fruit trees consist of at least two cultivars growing together. One cultivar provides the root system of the tree (the **rootstock** or stock) while the other provides the trunk, shoots and fruiting part (the **scion**). Propagation in the nursery consists of multiplying and growing the stock to the appropriate size and quality. The scion is taken from a very carefully selected source which should be (like the stock) virus free. The scion is attached to the root-stock by either budding or grafting. The combined tree is normally grown on for at least one more season before being transplanted into the permanent orchard site.

The scion and stock must be genetically compatible otherwise the bud or graft will not take. In most cases the scion and the stock must be from the same species but even then there may be some incompatibility problems. There are some successful unions between scion and stock from different species e.g. plum rootstock for peaches, and *Prunus mazzard* and *Prunus mahaleb* as rootstocks for the sweet cherry *Prunus avium*.

An interstem has been used in the past especially to reduce vegetative growth. The interstem is a stem grafted or budded onto the stock onto which a third scion is later joined. This adds to the cost of the tree.

Interstems were also used in situations where the desired scion and stock were incompatible. Interstems are not commonly used now because

- of the cost
- research has produced many more suitable stocks
- the interstem was commonly a weak point on the tree causing tree breakage under wind or crop load.

When a grower is planning for the development of a new orchard block he must choose very carefully the scion–stock combination. This choice will then determine the maximum orchard tree density, the range of possible training systems and whether tree support is needed.

The choice of a specific rootstock may be made for a range of possible reasons:

- The stock determines maximum potential tree size (an orchard of dwarfed trees must use an appropriate stock to control the tree to the size required).
- The stock has a major influence on tree shape especially natural branch angles. For instance Colt rootstock in sweet cherries produces broader branch angles than other stocks.
- The stock influences tree precocity, that is, how easily the tree develops fruiting behaviour (dwarfing stocks produce more precocious trees).
- A particular stock may be chosen to avoid an incompatibility problem (e.g. Colt rootstock in sweet cherries instead of Mahaleb for varieties such as Van).

- Some stocks tolerate wet or poorly drained soils.
- Some stocks are tolerant of acid soil conditions.
- Some stocks are drought resistant.
- Some stocks are tolerant of saline soils.
- Disease resistance (e.g. Mazzard F12/1 cherry is resistant to bacterial canker).
- Pest resistance (e.g. Northern Spy apple stock is resistant to overwintering woolly aphid).
- Anchorage (which dwarfing stocks lack).

The great care required in choosing the stock is particularly needed when a high density orchard is planned. If the trees are to be two metres apart along the row then the fully grown tree will occupy two metres along the row, be two metres thick and two to three metres in height. The stock is the appropriate control to allow the tree to reach this size and no further. Other methods of tree size control (constant heavy pruning or the use of growth regulators) are, by comparison, too expensive, time consuming and unreliable.

The scion determines the fruit variety and its features (e.g. size, shape, colour, flavour, aroma, firmness, shelf life, ripening season, disease resistance or sensitivity and pest resistance or sensitivity). The scion also determines the growth characteristics (spur or lateral bearer, vigorous or weak grower, strength of apical dominance, pest/disease sensitivity or tolerance, strong or brittle wood etc.). The flowering and fruiting features are also determined by the scion. All deciduous fruit trees have specific chill and dormancy requirements.

Chill requirement

Deciduous trees evolved in a climate where the winter temperatures were low enough to put the plant at some risk of freezing of soft tissues such as leaves and meristems (the growth areas). To avoid this damage, in the autumn such trees withdrew most of the nutrients from the leaves into storage in the stems and roots. When the leaves left are largely cellulose, they are dropped from the tree. At the same time the shoot meristems are protected by scale leaves and become buds. These tissues can then tolerate temperatures down to about $-10°$, depending on tree variety.

The trigger for this sequence of events is the day length as measured by the leaves. For each variety, as the critical short day length is reached in the autumn, the dormancy trigger is switched on. The speed at which the leaves fall is determined by factors such as the crop recently harvested from the tree, tree health and ambient temperatures. The internal control of these events is a plant hormone called abscisic acid.

The tree remains dormant as long as its genetic memory of winter. The control of this is the amount of abscisic acid inside the tree: the more abscisic acid, the longer the dormancy.

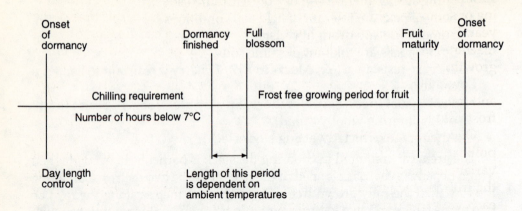

Figure 1.2 *Annual growth cycle of deciduous fruit trees*

Dormancy is finished when the abscisic acid no longer controls the inner workings of the tree. The abscisic acid decays in cold weather (under about 7°C). The more abscisic acid there is in the tree the longer is the period of cold needed to decay it. This period of cold is called the chill requirement. For Red Delicious apples the chill requirement is about 1500 hours. Late season cherries require from 1000 to about 1300 hours depending on the variety. Low chill fruit especially peaches and nectarines have chill requirements from about 40 to about 400 hours. High chill peaches require up to 1200 hours.

The chill requirement of a variety of deciduous fruit clearly is one factor which determines appropriate areas to grow it in. A variety should be grown in an area where its chill requirement is met in all but exceptional years. If the chill requirement is not satisfied the tree will blossom poorly and over an extended time. If the chill is a long way short of requirement the tree will not blossom at all.

As the abscisic acid declines approaching the end of the chill requirement, the tree stirs in activity. Usually the stock has a lower chill requirement than the scion, so the roots start growing as soon as soil temperatures allow. Easily the most sensitive parts of the scion to abscisic acid are the fruit (flower) buds. If their chill requirement if not met then the tree flowers poorly or over a long perod if at all. If the chill requirement is met then the flower buds swell as the inflorescence develops and are ready to open whenever temperatures allow. Once the chill requirement is met it is the ambient temperature only which determines when flowering will occur.

One of the major factors which determine where a particular variety of deciduous fruit can be grown commercially is that blossoming should occur during a period of low frost risk. For example a low chill variety grown in cold parts of inland New South Wales, such as Batlow or Orange, will have its chill requirement met by the end of June. A few days of Indian

14

summer and the trees will be in full blossom. The risk of a killing frost during or soon after blossom is extremely high in such areas at this time of the year. Frost control strategies such as the use of overhead misting or wind machines may not be adequate to protect the crop or even early vegetative growth.

Low chill fruit should be grown in an area where the winter is cold enough and long enough to provide adequate chill but where winter (and hence frost risk) is over when flowering begins.

Once flowering has occurred and all other requirements are met (e.g. pollination) fertilisation occurs and the fruit starts to develop. Each fruit variety has its own specific requirement for a frost free growing period for the fruit to reach maturity. This is another factor which determines where each variety of fruit must be grown. Red Delicious needs about 155 frost free days, Granny Smith needs about 180 and Pink Lady about 210 days. There would be some risk that Pink Lady may not mature at Batlow or Orange because of their shorter summers. Lady William which requires about 235 frost free days certainly would not mature in these areas in average years. An area with a much longer growing period is needed for this apple.

Other factors outside the scope of this book also influence the selection of the most appropriate areas to grow some fruits. For example nashi fruit do not fit neatly into the chilling requirement model and some fruits (e.g. peaches and cherries) require minimum levels of "heat units" to grow optimum quality. It is necessary however to consider how the trees must be managed to allow this simple model of annual growth cycle to be achieved. For example flowering will only occur if there are healthy flower buds on the tree. How does a grower manage his trees to produce adequate flower bud development?

Flower bud development

Flower bud development is dependent on a number of factors most of which can be managed by the grower.

Flower buds are initiated in the growing season prior to that in which the flowers will develop. In apples the major period of flower bud initiation is immediately after flowering. If there are far too many flowers at blossom time drawing on the limited amount of food, energy and building materials stored in the stems, the flowers will be weak and there may be pollination problems. There will be a relatively low percentage of flowers which set fruit. However because of the very large amount of blossom an excess crop of fruit will set. Heavy pruning prior to flowering will greatly reduce the number of flowers on the tree drawing on the same store of food. The percentage of flowers which will set fruit will be greatly increased because each is stronger. The percentage of flowers set is however much more predictable under these conditions.

When a flower is fertilised one result is rapid production of plant hormones. One group of these, the gibberellins, will inhibit the initiation of flower buds for the following season. So a heavy crop this year means a light crop next year. The answer to this problem in apples and pears is to chemically thin at blossom time and shortly after. This aborts most of the weaker flowers; they do not produce gibberellins and new flower bud initiation is improved. Follow up hand thinning may also be necessary to maximise this process as well as to maximise fruit size and quality.

This problem is not quite as important in stone fruits because the production of new vegetative growth this season has a big influence on flower bud initiation for next year. However, excess fruit set is still a problem in stone fruit. Because we have no reliable chemical thinners for these fruits at this time, removing much of the excess fruit bud during pruning and following up with early hand or mechanical thinning of the fruit is necessary.

The second major factor which influences flower bud initiation is the health of the tree throughout the previous season. If the tree is healthy, has access to adequate water and nutrients, is not stressed by excess crop, extreme temperatures or other adverse conditions and carries no significant pest or disease load, then initiation of flower buds is encouraged. While this is true of pome fruits it is even more important in stone fruits.

The third major factor which influences flower bud initiation is tree training. This is discussed in the chapter on apical dominance.

Fruit trees differ very widely in terms of where and when the fruit buds are developed. For example crop plants such as grapes (and roses as well) must be pruned very heavily every season. Only a perennial supportive framework is left as all last season's shoots are removed back to a stub with one bud. In the spring the plant grows back what was cut off. As the new shoot develops it produces leaves on the new growth and it also produces flower buds and then flowers. The fruit develops from these flowers, the fruit is harvested in the autumn and in the winter all the season's growth is pruned off and the cycle begins again.

With peaches (and nectarines, which are a hairless variety of peach)

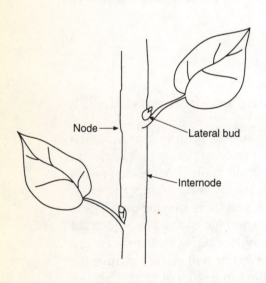

Figure 1.3 *Lateral bud in the axil of a leaf*

Node →

Lateral bud

Internode

16

heavy winter pruning encourages new growth in the spring. As these shoots develop at each node one or more buds also develop. There is usually a vegetative bud at each node and at many nodes one or two flower buds as well on each side of the vegetative bud. These buds will be found in the axil of each leaf on the current seasons growth.

In the next season the vegetative bud will develop leaves or a new shoot. From the flower bud will develop an

Figure 1.4 *Buds on a peach tree*

inflorescence or bunch of flowers. If the tree is stressed or very heavy pruning occurs, flower buds may degenerate into vegetative buds (particularly near to the pruning cut). But in most cases vegetative buds produce shoots and leaves and flower buds produce inflorescences. Clearly with peaches it is necessary to recognise the flower buds so that an assessment of potential crop can be made during pruning and the pruning strategy adjusted. In most fruit trees, flower buds are slightly longer and very much fatter than vegetative buds.

In peaches the season's growth produces new shoots and new flower buds. These flower buds will develop flowers in the following spring. So, unlike grapes where production of the flower buds and then the flowers occurs in one season, peaches produce the flower buds during one season and the flowers develop the following season. Buds in peaches are, in general, available for use only once. The buds are not perennial. Part of the skill in pruning peaches is to make maximum use of the buds which are available.

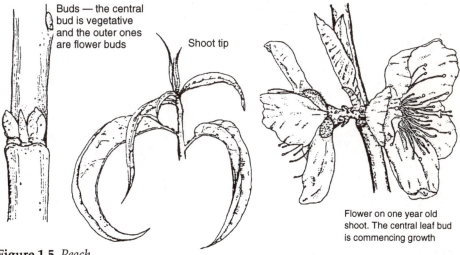

Buds — the central bud is vegetative and the outer ones are flower buds

Shoot tip

Flower on one year old shoot. The central leaf bud is commencing growth

Figure 1.5 *Peach*

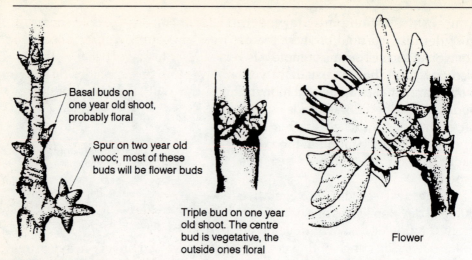

Basal buds on
one year old shoot,
probably floral

Spur on two year old
wooc; most of these
buds will be flower buds

Triple bud on one year
old shoot. The centre
bud is vegetative, the
outside ones floral

Flower

Figure 1.6 *Apricot*

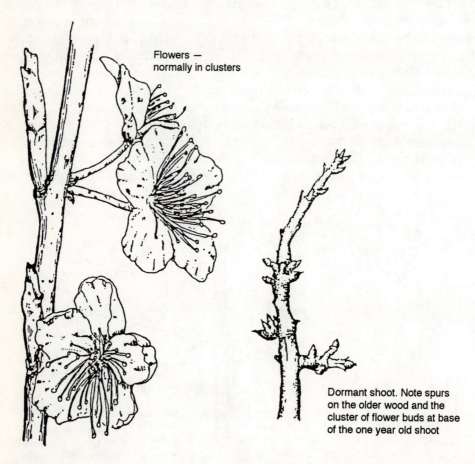

Flowers —
normally in clusters

Dormant shoot. Note spurs
on the older wood and the
cluster of flower buds at base
of the one year old shoot

Figure 1.7 *Plum*

In apricots flower buds are produced laterally on one year old shoots or in colder districts, and on more mature trees, on spurs on older wood. On young shoots the bud arrangement is very similar to that of peaches, that is, a central vegetative bud, usually with a flower bud on each side. On spurs (which are small shoots which have run out of puff), a number of flower buds are grouped together.

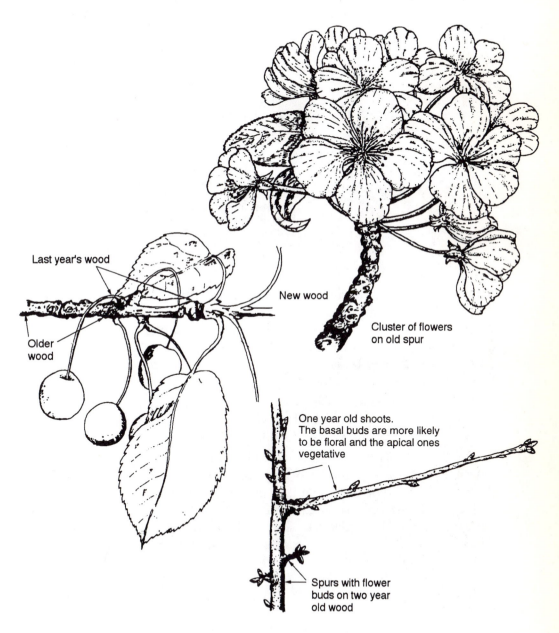

Last year's wood

New wood

Older wood

Cluster of flowers on old spur

One year old shoots. The basal buds are more likely to be floral and the apical ones vegetative

Spurs with flower buds on two year old wood

Figure 1.8 *Buds on a cherry tree*

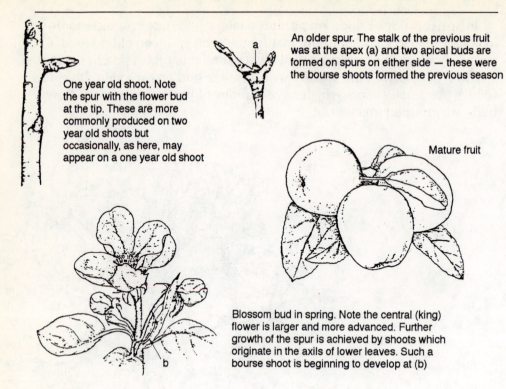

One year old shoot. Note the spur with the flower bud at the tip. These are more commonly produced on two year old shoots but occasionally, as here, may appear on a one year old shoot

An older spur. The stalk of the previous fruit was at the apex (a) and two apical buds are formed on spurs on either side — these were the bourse shoots formed the previous season

Mature fruit

Blossom bud in spring. Note the central (king) flower is larger and more advanced. Further growth of the spur is achieved by shoots which originate in the axils of lower leaves. Such a bourse shoot is beginning to develop at (b)

Figure 1.9 *Apple*

With plums (and apricots to a degree) from the nodes only vegetative buds develop. In the next season from some of these buds short shoots develop with a number of flower buds. The flower buds are not as clearly larger than those of peaches, but the short shoot or spur indicates the presence of many flower buds. These flower buds develop into flowers in the following spring. So the sequence with plums is:

– shoots with vegetative buds only in year one
– development of flower buds in year two
–development of flowers and fruits in year three.

The buds on a plum tree are available for use only once. Good pruning practice makes maximum use of the buds which are available. In apricots the buds are more persistent and are available for use over several seasons.

Apples and pears also have two types of bud. There are the vegetative buds which can produce leaves and/or shoots. Then there are mixed buds which produce leaves and an inflorescence. Mixed buds are clearly longer and much fatter than vegetative buds. Mixed buds do not occur in stone fruit. Recognition of mixed buds allows for an assessment of crop potential at pruning time. Good tree management requires that when inspection indicates a very high density of fruit bud on an apple tree the opportunity should be taken during pruning to remove much of the excess bud by pruning it off while also achieving the other aims of pruning.

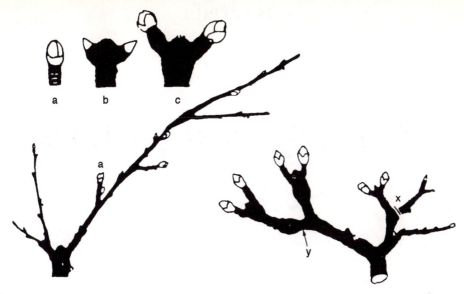

Figure 1.10 *Apple tree showing spur development:* **b** *and* **c** *show the development of* **a** *and* **x** *and* **y** *show spur pruning points*

In all fruit trees the development of flower buds is a prerequisite for cropping.

In simple terms the tree must not be under any major stress during the period of time in which flower buds normally develop. Stress can be caused by pests, disease, drought, nutrient deficiency, nutrient excess or imbalance, excess crop, waterlogging and other extreme environmental factors.

Vigorous rootstocks produce vegetative vigour in trees. Such trees are commonly difficult to persuade to crop and may be growing for ten or fifteen years before cropping commences. This situation is often exacerbated by irrigation and fertilising practices and inappropriate pruning strategies.

A range of tree management strategies can contribute to the development of flower bud and pruning and tree training can be used to manage the flower bud development on fruit trees.

CHAPTER 2

Pruning equipment

There is a very wide range of pruning equipment available. Some care must be taken to choose equipment which is suitable for the particular job. Modern pruning equipment and pruning aids allow for the speeding up of the pruning process and more efficient use of labour. The equipment chosen must be matched to the job required of it and must be used correctly. It must be appropriately serviced and maintained for maximum quality and output.

It is very strongly recommended that high quality pruning equipment be chosen. In general name brands should be selected as there is a number of cheap makes on the market of poor design and poor blade steel quality.

Secateurs

Secateurs are the most common and most basic pruning equipment. There are two main design types.

Anvil

The anvil type has a cutting blade which is sharpened from both sides and closes on a soft metal (usually brass) flat anvil. This type is very popular especially for the winter pruning of deciduous fruit plants where most of the cutting is of dormant material. Softer shoots may be excessively crushed. The best known quality brand is Rolcut. Ryset is a newer brand which is finding industry acceptance.

Bypassing or scissors

In this type the cutting blade is flat on one side and is sharpened from the other. The flat side of the blade passes beside the anvil like a guillotine.This

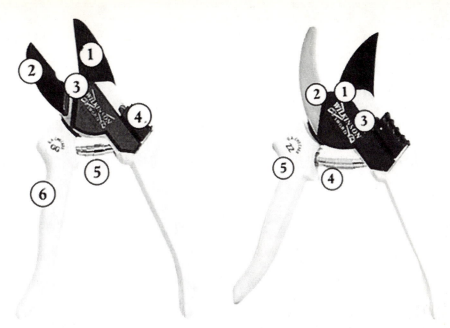

Figure 2.1 (a) *Anvil action secateurs.*
1. *Teflon coated blade and blade edge.*
2. *Sap groove.*
3. *Notch for cutting garden wire.*
4. *Opening/closing catch.*
5. *Spring.*
6. *Non-slip handles*

Figure 2.1 (b) *Bypass action secateurs.*
1. *Teflon coated blade and blade edge.*
2. *Notch for cutting garden wire.*
3. *Opening/closing catch.*
4. *Spring.*
5. *Non-slip handles*
(Wikinson Sword)

Figure 2.2 *Typical secateurs (Felco)*

type is usually preferred in propagation work or where soft tissue is to be pruned, e.g. in summer pruning, and is very widely used. The industry standard is the Felco brand which has a large number of model types to suit a wide range of applications.

In recent years several other brand names are finding acceptance. These include Wilkinson Sword, Ryset and Sandvik.

There is a wide range of styles of secateurs reflecting the large range of uses. For example there are left hand and right hand models, larger capacity and smaller capacity units, shock reduction models and models with a rotating handle. Select the model which suits your most common usage. No one model is "better" than another.

Secateurs are most suitable when the majority of the pruning is of shoots up to about 20 mm in diameter. Where larger cuts are required equipment with larger capacity must be used. Most cutting should be done with the throat of the secateurs and should be achieved with no twisting or sawing. Best results are achieved when cutting at right angles to the axis of the stem. This is especially so with sloping cuts.

A suitable pocket sharpening stone should be carried while pruning with secateurs and the blade lightly sharpened every four to five hours. The pivot point should be precisely tensioned as necessary and lubricated daily. The control springs should also be lightly oiled weekly. Replacement blade, anvil, pivot pins and springs are usually readily available.

Loppers

Loppers are essentially long handled, large capacity secateurs. They are being increasingly used in commercial orchards where larger cuts (more that about 20 mm and up to about 35 mm) are common. Loppers have three advantages over secateurs. Firstly they allow greater reach, secondly they permit larger cuts and thirdly there is less effort needed to make larger cuts. As with secateurs there are two main types anvil and bypassing.

Servicing loppers is very similar to servicing secateurs, with possibly a little more emphasis on pivot tension and pivot lubrication. Some makes have thin wall metal pipe handles and these are very easily bent and broken if too much force is used. Do not persist with attempting to use loppers when larger capacity equipment is more appropriate.

Hand saws

Where the required pruning cuts are more than about 30–35 mm or where it is unsuitable to use a lopper or secateurs, the most commonly used equipment is a hand saw. There are three main types.

Figure 2.3 *Loppers*
(Wilkinson Sword)

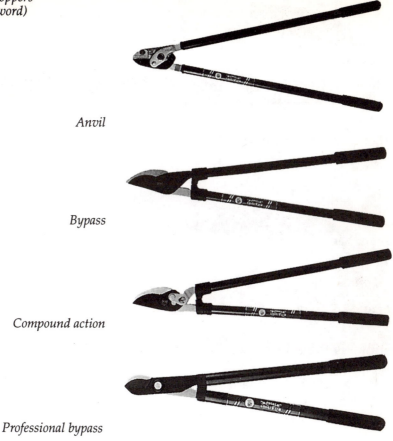

Anvil

Bypass

Compound action

Professional bypass

Bow

These are simple saws usually with a pipe metal frame either triangular or rectangular in configuration. Some makes have a rectangular solid steel frame with a wooden handle. The saw blade attaches easily on two pins and is tensioned by lever action or by a wing nut. The main disadvantage of this type of saw is its size in a cramped area within a tree. For this situation thicker saw blades are attached to a simple wooden handle.

A feature of all of bow saws is that the saw teeth are filed and alternately bent. They can be resharpened and the teeth reset.

Blade

These have come onto the market in recent years and proved very popular. The blade is attached usually to a plastic handle and can be carried in a belt sheath. Some styles can be folded. The distinguishing feature of this type of saw is that the high class (if somewhat brittle) steel blade is taper ground and each tooth is honed and not bent. It produces a remarkably clean cut which aids in fast wound healing. In contrast to most other saws the cutting

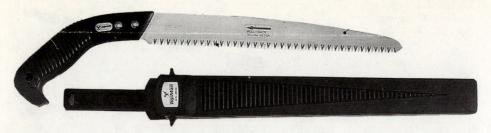

Figure 2.4 *Blade saw and sheath (Topman)*

Figure 2.5 *Folding blade saw (Topman)*

stroke is the pull stroke. This requires a small modification of cutting technique.

The industry standard is the ARS Japanese saw and a similar unit is the Topman saw which is also of Japanese manufacture. For a small saw (blade length 300 mm) they do an excellent job on cuts of up to about 250 mm.

Pole
These have a small blade mounted on a long pole so that saw cuts can be made from the ground on branches up to about 4 m above ground level.

Chain saws

At one time it was considered almost sacrilege to use a chain saw on a fruit tree. Nowadays with the increasing understanding of the requirements of pruning and the need for quick and cost effective pruning, they are more generally accepted.

The appropriate use of a chain saw is the removal of whole limbs or major branches from a tree. This may be to overcome a major shading problem or to remedy a structural fault within the tree. In a modern orchard especially those on semi-dwarfing or more vigorous rootstocks, a walk through the orchard with a chain saw is an appropriate way to start the pruning season. As discussed later, always make the major cuts first. A chain saw is likely to be the most efficient way of doing this.

Figure 2.6 *Air compressor on a trolley (Campagnola)*

Fluid power pruners

Very widely used on commercial orchards are the power pruners, especially pneumatics.

Pneumatic pruners require a suitable air compressor unit which may be:
- on a trolley (e.g. Campagnola brand)
- on a tractor as a power take off unit
- on a trailer or elevated pruning platform
- on a dedicated mobile ladder unit (e.g. Afron brand).

Easily the most common pneumatic pruners are air powered secateurs. They are much faster and far less tiring than conventional secateurs or loppers. Because the effort needed for major cuts is only the pressing of the air trigger, most people prune better with pneumatic equipment. This is because cuts which would be difficult with secateurs or loppers are easier with power pruners. The pruner then has no hesitation in performing the appropriate cut.

Disadvantages

Cost

Compressor units can cost about $1,000 to $3,000 and handpieces cost from about $150 to $450. In addition a water separator and an oil injection unit are required. Water vapour is condensed during compression of the air and must be separated before reaching the pruner.

27

Figure 2.7 *Compressor for a tractorpower take off (Campagnola)*

Because of the much higher work rate and the extra pressure on pivot points, automatic oil lubrication is essential. A very efficient way of achieving this is by a unit which injects a fine mist of light oil into the pressurised airstream.

Danger

The risk to the pruner is higher particularly where there is lack of suitable experience. The jaws of the pruner close as soon as the control is pressed whether a finger is in the way or not. There have been many cases reported of finger or hand damage. The pruner must develop a technique when using pneumatic equipment to minimise the risk. For example, it is important to concentrate on keeping the other hand well away from the cutting site especially when using it to pull prunings away. A reflex action when attempting to clear a tangle is to squeeze the trigger. This also happens when the hose becomes tangled in the tree or prunings. It is also very common to cut the air hose.

Equipment failure

If the compressor or handpiece fails you are out of business. A priority should be to service the compressor at least twice a day. This usually consists of checking the oil level in the compressor and checking the alignment and tension of any vee belts. A second unit or handpiece is advisable as well as appropriate spare parts. If all else fails you may need to revert to hand pruning with old fashioned secateurs.

Figure 2.8 *Pneumatic pruners on an Afron pruning platform*

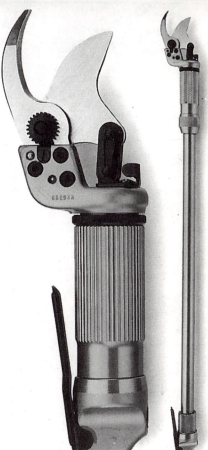

Figure 2.9 *Air powered secateurs (Felco)*

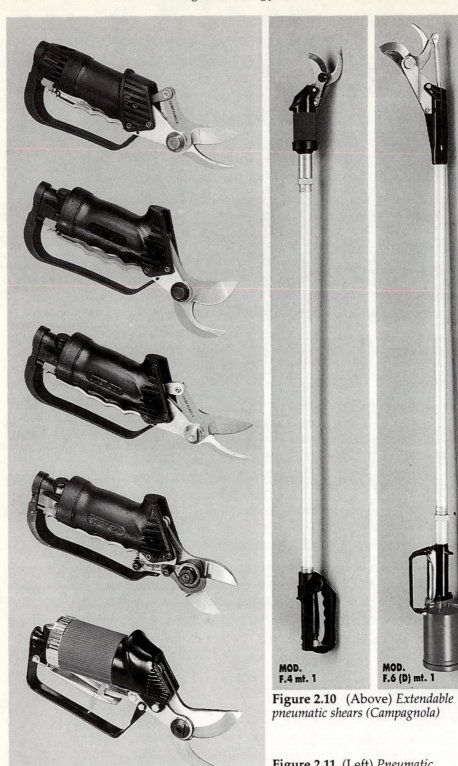

MOD.
F.4 mt. 1

MOD.
F.6 (D) mt. 1

Figure 2.10 (Above) *Extendable pneumatic shears (Campagnola)*

Figure 2.11 (Left) *Pneumatic pruning shears (Campagnola)*

Figure 2.12 *Hedging machine in an apple orchard*

Advantages

The advantages of pneumatic pruners are that they are very fast, very efficient and far less tiring than conventional equipment. Because of the lack of physical effort in heavy cutting, most people prune better with power pruners. They will not hesitate to make an appropriate heavy cut with power pruners. People tend to select easier cutting when using hand pruners.

Other pneumatic pruning equipment includes:

- a hand unit on an extension pipe (about one metre)
- a pneumatic circular saw on an extension pipe
- a pneumatic chain saw on an extension pipe

Hydraulic pruning equipment is also available but is not commonly used in orchards. There is some use in the nut industry and for urban tree management. The equipment works well but for most tree fruit crops it is very heavy and the hoses are hard to manage. Very large capacity hydraulic loppers, chain saws and circular saws are available.

Hedging machines

Hedging machines most commonly consist of a bank of three to ten large circular saws mounted on a hydraulic boom. The boom can be placed in a vertical orientation, a horizontal orientation (at a range of heights) or at any angle in between, using hydraulic power. The hydraulic power also operates

Figure 2.13 *Sickle bar hedging machine in a dwarf apple orchard*

the circular saws. The equipment is usually mounted on a dedicated four wheel drive tractor of 100 kw or greater power. The tractor is fitted with a very strong protective cage to protect the driver from high speed pruning offcuts from the saw blades.

Hedging machines can remove a lot of timber from trees at a very fast rate. A typical machine can cover up to three hectares per hour per cutting orientation. Such machines are commonly used to reduce the tree height of a block, to summer prune (especially peaches and nectarines) and to help control tree shape. One of the reasons for the design of the Tatura trellis system (see the chapter on training systems) was to allow for mechanised pruning both on the inside and the outside of the vee shaped trees, using such equipment. Hedging machines are commonly used where a large block of trees is in urgent need of removal of excess shoots which are carrying excess fruit bud or are causing major shading problems. Commonly there is a shortage of time or suitable skilled pruners.

Another type of hedging machine uses a sickle bar mower similar to that used in harvesting lucerne and similar crops. This machine is also useful for summer pruning of stone fruits and is less severe on the tree and fruit. A machine like this is a must for Lincoln Canopy systems and is useful in other hedged or trellis systems for both summer and winter pruning.

Trees which have been mechanically hedged need follow up detailed hand pruning with power pruners in one or two seasons.

Disadvantages of hedging machines include rougher wound cuts which heal slowly and may be an entry point for bacterial or internal woodrot fungal diseases. Hedging can induce vigorous and uncontrolled regrowth unless treatments such as NAA (Naphthalene Acetic Acid) paints or sprays are used.

CHAPTER 3

Apical dominance

Know your fruit tree

This chapter is the most important in this book. It is my very strong view that learning to prune fruit trees is best achieved by learning how they grow. The knowledge of how tree growth occurs, how it is managed and controlled within the tree and how the tree will react to a pruning cut is critical. Pruning cannot be learned by following a template process e.g. cut each young shoot back to four buds (or some similar description). A tree is pruned because the tree needs pruning. If it does not need pruning, then do not prune it.

Pruning is only one tree management tool. There are many others. Pruning is used as an integral part of the tree management process. It is used to assist the tree's natural growth behaviour to develop an appropriate tree framework and then to produce early and consistent crops of quality fruit.

Growth behaviour

Trees like animals are complex biological organisms. Consider the immense range of biological activities we hear about in medical science. The range of activities in plants is no less.

The growth, development and behaviour of fruit trees is the result of the balance of all the growth factors which have an influence on the tree. The most important balances in which we are interested are the balances of the naturally occurring hormones in the tree and the balances of the plant foods and nutrients.

In animals and humans we are aware of the influence of hormones in the growth and development of the individual. A hormone is a naturally occurring chemical produced in very small quantities in one part in the body. It is translocated to target tissues or organs where, at extremely low concentrations, it triggers and controls a growth or developmental response.

In animals there are many types of hormones produced in different glands and with different target organs. In plants there are several types of hormone as well. The most important two groups for our discussion at this point are the auxins and the gibberellins. Synthetic versions or analogs of these are marketed as plant growth regulators. These two plant hormones are involved in controlling the growth of trees. This is especially important in the spring when they control the growth response known as apical dominance.

An understanding of apical dominance is vital in the understanding of the pruning and training requirements of the fruit tree. Most examples of poor orchard design and management, of badly pruned or poorly trained trees can be traced back to a lack of this understanding.

As the tree comes out of dormancy in the spring the terminal (or apical) bud in the shoot is triggered into action and it commences the complex process of spring growth. Inside the bud is a tissue or group of cells known as a **meristem**. Meristems in plants are where growth is produced by the process of cell division producing new cells. As this process continues new cells are being continually added to the plant, producing the increases in length and thickness of the plant's structure known as growth.

The most active meristems in plants are those in the tip of the shoots (shoot apical meristems) and the tips of the roots (root apical meristems). Equally important is the intercalary meristem commonly known as the **cambium**. This is the layer of cells which separates the bark (containing the food conducting tissue called the **phloem**) from the wood which contains the water and nutrient pumping tissue (known as the **xylem**). The cambium is the meristem which produces the new cells which increase the diameter of the stem.

Every bud contains an active or potentially active meristem. These buds are often targeted in the pruning process to be triggered into activity to produce new growth at a point and in a required direction. This means that the pruner can select the meristem which is to be activated. There are also meristems in the roots which produce branch roots as the plant grows.

When the shoot apical meristem commences activity in the spring, cell division occurs which adds new cells to the length of the shoots which then elongate to their optimum size. This produces the very obvious shoot elongation of spring. At the same time meristematic activity in the cambium is producing new layers of phloem cells (in the bark) and new xylem cells (added onto the existing wood). Under the ground, roots are increasing in length and diameter for the same reason and branch roots are developing.

As this activity begins in the shoot apical meristem it produces a hormone, a chemical known as indole-acetic-acid (I.A.) which is a naturally occurring member of the group of hormones known as auxins. Many synthetic auxins,

34

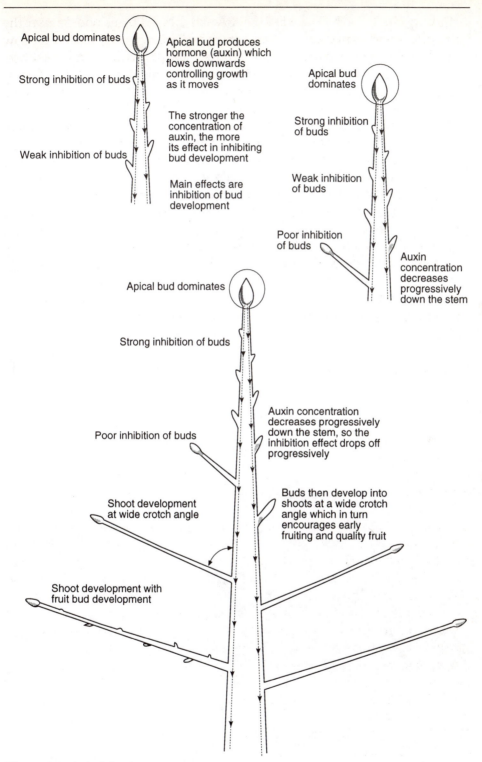

Apical bud dominates

Apical bud produces hormone (auxin) which flows downwards controlling growth as it moves

Strong inhibition of buds

The stronger the concentration of auxin, the more its effect in inhibiting bud development

Weak inhibition of buds

Main effects are inhibition of bud development

Apical bud dominates

Strong inhibition of buds

Weak inhibition of buds

Poor inhibition of buds

Auxin concentration decreases progressively down the stem

Apical bud dominates

Strong inhibition of buds

Poor inhibition of buds

Auxin concentration decreases progressively down the stem, so the inhibition effect drops off progressively

Shoot development at wide crotch angle

Buds then develop into shoots at a wide crotch angle which in turn encourages early fruiting and quality fruit

Shoot development with fruit bud development

Figure 3.1 *Apical dominance*

including 2,4-D, 2,4,5-TP, 2,4,5-T, N.A.A. and N.A.D. are widely used in horticulture and agriculture for a range of purposes.

In the stem the auxin moves downwards in the phloem, controlling many growth features as it does. For example it is instrumental in controlling the elongation of the new cells produced at the shoot tip. Auxins are also involved in the production and pattern of placement of the new buds on the nodes of the new shoot as it grows. More importantly from a tree training point of view, the auxin inhibits the newly developed buds from producing anything other than leaves. The apical bud or meristem is using the auxin to dominate the shoot growth and preventing any side shoot from developing and competing. The stronger the apical meristem, the higher the concentration of auxin and the more there is of this dominance. This is called apical dominance.

The auxin concentration declines during its flow down the stem. There are a number of reasons for this. Some is being used up by the bud inhibition process and the cell elongation process. Some is being used up by other chemical reactions in which it is involved and some is being destroyed by contact with light. At the same time as all this has been going on, other

Figure 3.2 *Unpruned cherry trees showing apical dominance growth*

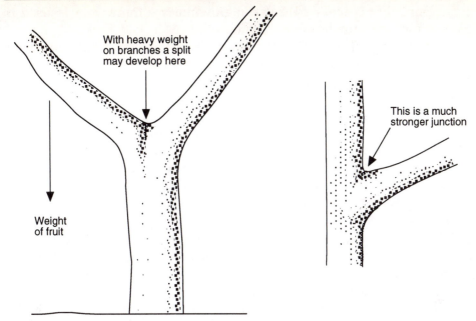

Figure 3.4 *Strong branch junction (branch angle)*

hormones such as cytokinins and gibberellins are being produced within the plant. Especially important for tree training purposes are the gibberellins which are largely produced in the roots and travel upwards inside the plant. The result is that at various points within the plant there are varying balances between the hormones.

The most important balance is that achieved at a point in the stem where a critical balance of auxins and gibberellins is achieved. At this point the auxin cannot quite inhibit the bud. The bud produces a shoot with a most important feature. Because of the hormone balance situation the shoot will grow out at a broad angle to the stem. This broad angle is critical in tree training because not only is it a strong angle (to support the weight of a crop of fruit) but is precisely the branch angle which initiates the induction of fruit bud and the subsequent production of quality fruit. This shoot with its own developing apical meristem continues to grow at an angle determined by the everchanging hormone balance within.

Effect of pruning and training

What does all this mean in tree training terms?

- By leaving the apical meristem on the tip of the shoot (i.e. by **not** heading the shoot during pruning) the maximum shoot elongation and maximum diameter increase (from the auxin encouraged cambium) is achieved. You can grow the tree framework faster and stronger by not heading the shoot.

- By leaving the apical meristem on, shoot side branching is achieved some distance below the shoot tip at a broad branch angle. This is the strongest possible connection to the main central stem.
- The sideways growth of the side shoots produces the optimum tree shape for light penetration.
- The sideways growth will be about 60 degrees from the vertical which is the optimum branch angle to initiate the production of fruit bud.
- This angle of side growth is also the optimum to achieve the balance between fruit production and the amount of vegetative growth needed to produce optimum fruit quality.

The meristem at the tip of the developing side shoot is just like the meristem at the tip of the main leader of the tree, with one very significant difference. The shoot is not vertical but growing at some broad angle to the main shoot. This means that this meristem does not have as much dominance as the apical meristem. It is also being influenced by a higher concentration of gibberellins than the apical meristem because being lower in the tree, it is nearer to the source. For a side shoot this means that the inhibition process reduces further branching from the buds as they are produced by growth. But because the hormone balance is different from that in the main shoot or leader the buds can easily develop as flower or mixed buds.

Figure 3.5 illustrates this point. Side shoots growing in the Zone A orientation will be very vigorous producers of shoot growth but have almost no ability to grow fruit buds. Side shoots growing in Zone B1 orientation are in the transition zone where fruit buds can be induced but the shoot is still dominated by shoot growth. In Zone B2 fruit bud development is easy to induce but at the same time there is enough vegetative growth to maintain vigour in the shoot and the production of abundant food for developing fruits. Such shoots commonly bend down into Zone B3 under the weight of the fruit, and their vegetative growth will be reduced.

Shoots originally in Zone B3

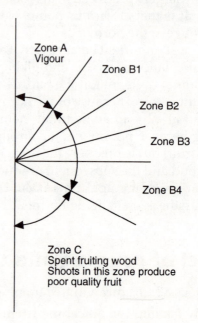

Figure 3.5 *Angles of growth of shoots*

are in a position where abundant fruit bud is easily produced and vegetative growth is still adequate. However, the dominance of the terminal bud is very reduced and excess small side shoots are now a possibility, creating too much shading. If the shoot bends into Zone B4 during fruit growth, elongation growth of the shoot ceases at the horizontal position and small side shoots oriented upwards will take over. This can largely be remedied during the next pruning season.

Shoots which are permanently in Zone B4 will still produce fruit but the quality of the fruit will be determined by the growth of upwards oriented shoots. This is because the terminal bud is no longer meristematic and will produce no elongation growth.

Shoots permanently in Zone C produce poor quality fruit because there will not be enough vegetative growth producing the food which the fruit needs. It is also highly likely that this wood is shaded. Such wood is of no use to the tree or grower and should appropriately pruned.

All this has been achieved by simply **not** pruning (heading) the leader (the main vertical shoot) of the tree. You have let the tree grow the way it wants to grow as this coincides with the most efficient tree shape. You have trained the tree by understanding the way it wants to grow and by making the decision not to prune. The result will be a strong central leader with whorls of sideways shoots (like spokes of a wheel) well below the tree's apical meristem. Apical dominance is thus used to control the tree's growth.

We can liken the situation to that in a political party where a very strong leader inhibits the potential competition from senior team members. This analogy will be used later to explain other facets of tree training using apical dominance.

The discussion so far has suggested that the concept of apical dominance and its use in tree training is very simple. To some extent this is true but there is more to the concept which we will now need to consider. For example, what are the factors which determine the strength of apical dominance? Clearly not all trees grow in the same way.

Factors affecting the strength of apical dominance

Several factors determine the strength of apical dominance:

(1) Every fruit tree species, variety and cultivar is genetically different from every other. This results in a very large range of growth behaviour between species and within one species. Peach trees have a different growth habit from pears and Granny Smith apple trees grow differently from a Jonathon. In terms of apical dominance we can make some broad generalisations:

- Pears have extremely strong apical dominance.
- Cherries have very strong apical dominance.
- Apples generally are also very strong but there is a big range.
- Plums have very strong dominance.
- Peaches have medium strength dominance which is stronger when the tree is young and weakens with age.

(2) The fruiting behaviour of the cultivar influences the strength of apical dominance. For example lateral bearing apples have very strong dominance but it is weaker in spur bearing varieties.

(3) The precocity (or ease with which the variety produces fruit bud early in the tree's life) of a variety is also important. For example a tree which is normally apically dominant will lose most if not all of the dominance if it is allowed to carry too much fruit too early in the tree's life. This commonly leads to stunted trees if the situation is not remedied.

(4) The verticality or erectness of the stem is a major factor controlling apical dominance. Dominance is greater and shoot growth is maximised if the shoot is vertical. This dominance weakens in direct proportion to the angle away from the vertical. In trees on moderate to vigorous rootstocks this vertical growth is relatively easy to achieve. In trees on dwarfing rootstocks the leader may have to be supported to maintain its vertical orientation. This is part of the reason for the need for tree support in high density orchards on dwarfing rootstocks.

(5) Leader competition reduces absolute dominance. For example if the leader of a tree is pruned back (or headed) the original dominant leader is lost. The auxin it produced is now missing so the auxin controlled inhibition of lower buds disappears. The buds closest to the point of pruning now attempt to become the new leader. Using our political party analogy, if the leader accidently falls in front of a bus, all of the previously inhibited major contenders now compete with each other to become the new leader.

Each of the new shoots grows vertically; each has an equal share of the growing resources. So where the tree had one strong vertical shoot it now has several growing from the highest remaining buds and growing vertically competing with each other.

The grower now has to make a decision. Are all new leaders kept so that a narrow vase tree results? Or is a central leader tree re-established by keeping the top shoot and cutting its competition off? Or is the tree trained but not pruned by keeping the top shoot as the new leader and spreading the others away from the vertical thus reducing their dominance? This last choice is often made in developing a central leader tree from a newly planted tree. That is, the tree is headed after planting, the best leader kept, and the rest spread. This is a critical move if the tree is a spur type or on a weak rootstock. It is called the head and spread method.

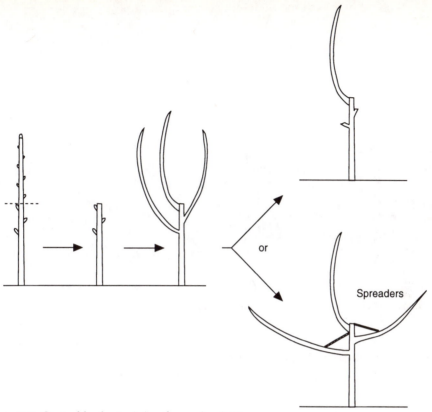

Figure 3.6 *Central leader training from a headed tree*

(6) The stronger the rootstock the stronger will be the dominance. The dwarfing rootstocks now available in apples (and some other fruits) greatly reduce dominance so other strategies must be implemented to maximise the dominance which remains. These include supporting the leader and spreading the other shoots.

(7) Normal tree health and vigour influence the strength of the dominance. A healthy tree with adequate nutrition and water will have maximum dominance.

(8) Any major stress on the tree will reduce dominance. These stresses may include pests, diseases, nutrient deficiencies or imbalances, excess crop load, heat, drought and other growth factors.

(9) Any damage to the apex of the tree can have a similar effect to heading back. The damage could be physical such as wind or hail damage. It could be disease such as curl leaf or shoot blight in peaches. It could be powdery mildew in apples. It could be shoot girdling by heliothis caterpillars or a moth borer. It may be picker or machine damage. Whatever the cause, the remedy is to select a new leader as soon as possible and manage the new leader.

Figure 3.7 *Spur-type Red Delicious apple tree showing the development of one leader and the spreading of the other vertical shoots*

(10) Elevation position is important. When a new leader is chosen for any reason, the choice must be based on choosing the best candidate possible. It should be a strong shoot pointing in the best direction. Once it is chosen it must grow from the highest point on the limb. Sometimes a weak shoot will grow from the top bud and a stronger shoot from lower down. Or the tree height may be reduced by choosing a new leader lower down. In both cases the best new leader available must grow from the top of the shoot.

(11) The imposed training method will influence strength of dominance. For example any trellised system which spreads leaders away from the vertical will reduce the strength of the dominance.

(12) The use of applied plant growth regulators or pesticides which have some

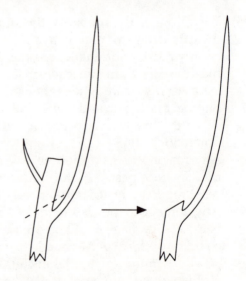

Figure 3.8 *Because of the nature of apical dominance a chosen leader must be the strongest available shoot and grow from the top of the limb*

growth regulator effects will influence dominance. Because dominance is controlled by the hormone balance within the tree, the application of any plant growth regulator may change this balance. The use of Cytolin (which contains a synthetic gibberellin and cytokinin) for example can overcome the auxin induced inhibition of buds and induce side branching where required.

Using apical dominance to manage the tree

Looking closer at the "ideal" central leader tree discussed earlier, we will observe that as soon as the side bud is no longer inhibited from developing by the downward flowing auxin, it develops a new shoot with its own leader and apical bud. The growth of this shoot and the development of the the apical bud is still influenced firstly by the hormone balance at the bud and then by the changing hormone balance created by the new developing leader. The strength of the dominance of this new leader is determined by its genetics and its verticality. This will vary from one cultivar to another.

If the side shoot becomes too dominant and too vertical it may start to compete with the original leader. Too large a proportion of the tree's vegetative growth is from the base of the tree rather than the top. This is called **basitonic growth** and it is typical of some apple cultivars such as Red Chief Red Delicious. The answer with a basitonic tree is to protect the main leader and to spread the side shoots to weaken their dominance. This

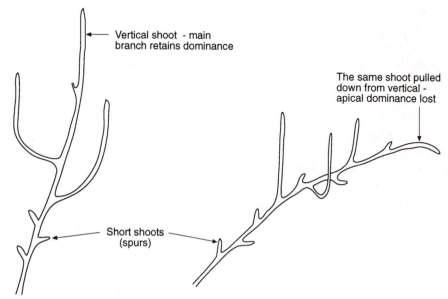

Figure 3.9 *Strong and weak apical dominance due to branch orientation*

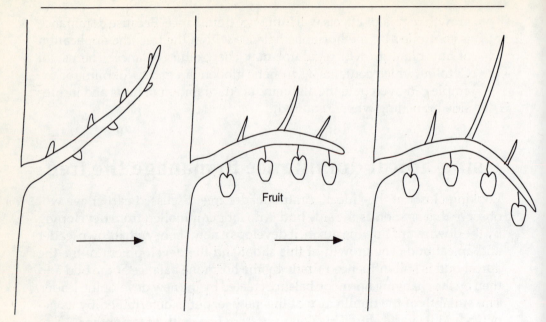

Figure 3.10 *Shoot growth behaviour when pulled below horizontal by the fruit load*

is the head and spread technique developed by Don Heineckie (previously from Washington State University).

Because of the relative weakness of the leader of the side shoot, fruit bud development is easy to initiate. So by not pruning but controlling the side shoots, trees can be persuaded very easily to crop. The cropping itself will then influence tree development. So spreading and cropping are two ways of training trees without pruning.

In the light of this concept we can examine more closely what happens to the side shoot if it is pulled down by the weight of the fruit to horizontal or even below.

As shown in Figure 3.10, when the side shoot bends under the weight of the fruit, the angle begins changing from the 60 degrees to the vertical towards 90 degrees from the vertical. At this angle (that is, horizontal) the apical meristem has lost its dominance powers. The shoot will not produce elongation growth. As the shoot bends below horizontal the bud at the tip of the shoot is no longer the dominant bud. The dominant bud is that one at the highest point on the bend of the shoot. Because the apical bud is producing no auxin this bud is not inhibited at all so it will begin to grow a shoot. If nothing else now changes this shoot will be vertical and very strong. This type of growth is commonly called a water shoot. Water shoots are a problem because they grow strongly and vertically inside the tree canopy producing crowding and shading.

The water shoot develops because apical dominance control has been lost on the side shoot as its apical bud drops below horizontal. If the side

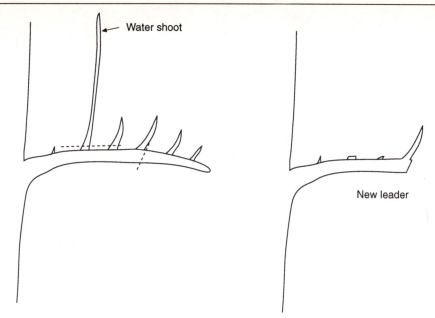

Water shoot

New leader

Figure 3.11 *Watershoot management*

shoot (especially the apical bud) has been prevented from dropping below horizontal then the water shoot is very unlikely to develop. If it does develop then clearly it must be removed completely.

If the side shoot continues to drop under the weight of the developing crop what happens is that as every bud on the bowed part of the shoot becomes the highest (hence dominant) bud, it develops a shoot. But before this shoot develops, another bud takes over as the new top bud and another weak shoot develops. The result is a bowed main side shoot with a number of short shoots developed on it. The correct remedial procedure is to prune the end from the side shoot back to the best positioned and oriented of the weak shoots (Figure 3.11) (about 60 degrees from the vertical), which will become the side shoot's new leader. This procedure may need to be repeated at intervals (Figure 3.12).

This means that the fruit bearing side shoots are largely renewed on a regular basis. But the renewal pruning must only be done when it is necessary. The overall long term aim is to maintain the fruiting arms (the side shoots) at an angle of about 30 to 60 degrees from the vertical. This angle is important because it is wide enough to prevent the side shoot ever competing with the tree's leader. The angle is also steep enough to allow for some strength of the apical bud so that it may impose its own apical dominance on that shoot. This allows for some continuing vegetative growth to retain the vigour of the shot which will in turn optimise fruit size and quality (Figure 3.13).

Similar sorts of problems can occur at the tip of the tree where the tree's

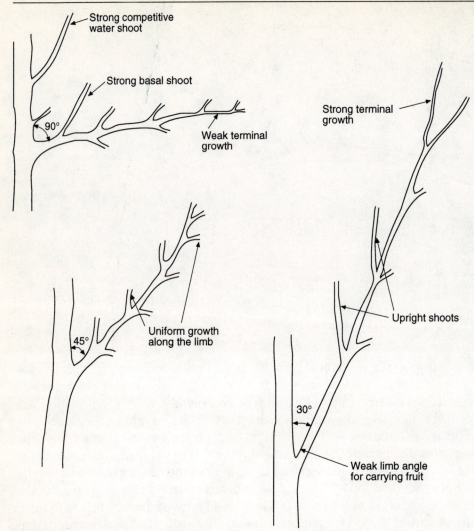

Strong competitive
water shoot

Strong basal shoot

90°

Weak terminal
growth

Strong terminal
growth

45°

Uniform growth
along the limb

Upright shoots

30°

Weak limb angle
for carrying fruit

Figure 3.12 *Limb angles can be used to reduce or encourage shoot growth fruitfulness*

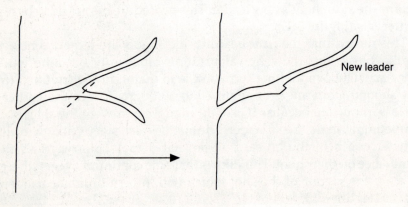

New leader

Figure 3.13 *Dormant pruning of low angle branches*

Figure 3.14 *Peach tree showing a side shoot which has been pulled down to horizontal by the weight of a crop of fruit*

Figure 3.15 *The peach tree in Figure 3.14 after pruning*

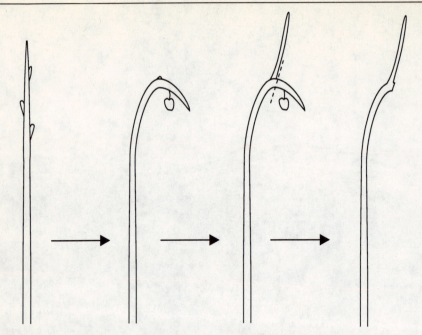

Figure 3.16 *Loss of apical dominance because of fruit weight. The final stage shows the effect of remedial pruning*

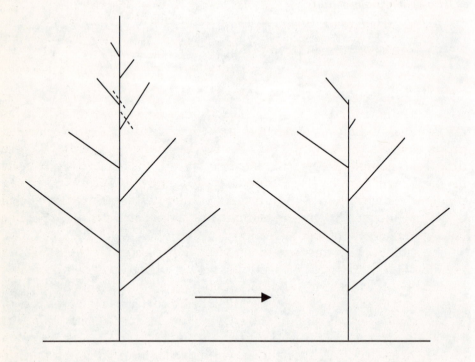

Figure 3.17 *Reduction of the height of the tree while maintaining apical dominance control*

Figure 3.18 *Central leader apple tree where the central leader has been encouraged to grow vertically for five years, controlling the tree's development. When the tree was large enough the leader was allowed to bend under the weight of the crop. As it bent and lost its dominance, another leader developed from lower in the tree. The tree's height has been controlled by training not pruning*

apical dominance should be strong. If the top of the shoot bends for any reason (a common one is fruit on a shoot which is too weak to carry it), as the terminal bud moves away from the vertical its dominance declines in proportion. As the terminal bud approaches horizontal its dominance disappears completely. The dominant bud now is the one at the top of the bend. So it will grow a vertical shoot. If the bending continues (as discussed previously) then several new vertical shoots may develop. To regain control of the tree the most suitable leader must be chosen and the top of the tree cut back to it. Any competing leaders must be removed (Figure 3.16).

The same concept can be used to reduce the height of the tree while still maintaining apical dominance control. Simply cut the top back to a suitable candidate for leader. This candidate should preferably be close to vertical, fairly strong and at the required height. Having cut back to it (i.e. electing it as the new leader) then all possible competing leaders must be removed (Figure 3.17).

Tree height control especially in central leader trees often occurs naturally when the leader is permitted to bear fruit, and bend below the horizontal; new vertical shoots develop and a new leader must be chosen.

It should now be clear that choosing an unheaded leader either at the top of the tree or on the side shoots which are the fruiting arms leaves the pruner in control of the tree. By constantly heading during the pruning process, apical dominance and control is lost and the regrowth will be uncontrolled. Heading always produces a number of shoots (the competing new leaders) increasing shading problems and increasing the pruning which will be needed in the next season. If heading back is needed then head back to a shoot no matter how weak. It will then dominate in the next season, producing growth in the direction you have chosen. Heading back may be necessary for a number of reasons but the most common is to reduce excess fruit bud or excess height/length. In either case head back to a newly elected leader which is pointing in the direction you want the regrowth to develop in.

CHAPTER 4

Reasons for pruning

A fruitgrower might be tempted to ask: "I go to a great deal of trouble and expense to irrigate, to fertilise, and to control weeds, pests and diseases, all to make the trees grow; why do I have to prune them back?" This is particularly a temptation when the pruning season is a mountain climate winter.

Pruning has four major uses which make it such an important component of fruit tree management:

1. tree training purposes
2. control of the fruiting process
3. maximising fruit quality
4. its contribution to pest and disease control.

Tree training

(a) No matter which training system has been chosen, the grower must decide where he wants the **scaffold** (or permanent framework) limbs to develop. So, appropriate leaders are chosen and their competition removed by pruning. This now allows for the apical dominance to quickly establish the tree's framework according to the chosen training system. This decision is made at planting when the tree is headed back.

(b) As the tree develops, the shape it will grow in is decided each pruning season. Apical dominance should be used for this development by **not** heading the required leaders and by **removing possible competing leaders**. Adjustments are made as the tree develops. It is not uncommon to find after four or five seasons that too many leaders have been selected and now the tree has become too crowded. It is also common to find that leaders

Figure 4.1 *One year old peach tree after pruning. The major scaffold limbs have been decided and left unheaded*

Figure 4.2 *One year old peach tree which has lost apical dominance during the last growing season (probably from Brown Rot or Leaf Curl fungus)*

have not grown precisely in the required direction. As a tree fills its allocated orchard space one branch may need to be reduced.

(c) Remember that a tree reacts to being pruned by trying to grow back what has been pruned off and it tries to do this as near to the point of pruning as possible. If a limb has been cut out because the tree was far too dense then the tree will try to regrow the limb in the next growing season.

One of the skills of pruning is to **direct the regrowth** which must come into an appropriate direction. This is achieved by cutting back to an appropriate mini-leader each time a major cut is made. This leader must be at the top of the remaining shoot and it must point into a space where regrowth will not produce excess shading. The amount of regrowth can also be reduced by selecting a near-horizontal leader where some of the regrowth can be absorbed as fruit and fruit-bud development.

(d) During the ongoing development of the tree more of the pruning will be used to **adjust the tree** rather than in the original development of the framework. Major fruiting arms will be chosen and encouraged. They will be spaced at suitable distances apart (about one metre). When apical dominance has finished its job of developing the tree's framework, the tree height may need to be reduced to a lower suitable leader.

(e) When the tree has filled its allocated **orchard volume** it will be necessary to keep the tree in its space. If one limb intrudes into another tree's space

Figure 4.4 *Four year old peach tree before pruning. The number and arrangement of the fruiting arms has to be determined*

Figure 4.3 *The peach tree in Figure 4.2 after pruning. The leader has been selected*

Figure 4.5 *The peach tree in Figure 4.4 after pruning. Note the direction and spacing of the fruiting arms*

way will need to be kept open so that tractors and other equipment are not impeded by tree growth.

(f) The trees framework must be established before the tree is permitted to crop heavily. One very effective way of **reducing precocious cropping** is to prune off excessive fruit bud early in the tree's life. This is especially important for stone fruit. The pruning can be used to enhance the framework development.

(g) Fruiting arms which:
- touch the ground under crop weight, and will pick up fungi from the soil and contaminate the fruit
- are in the way of cultivation equipment
- are in the spray pattern area of under-tree weed control units
- will be hit by orchard equipment in the traffic ways, or
- will be hit by the mower

are of no value so they should be pruned off.

53

(h) Fruiting arms need to be rejuvenated at intervals for maximum fruit quality. The best fruit grows on limbs which are growing. This is achieved by cutting back and regrowing the limbs. This is especially important in trees on very dwarfing rootstocks. When a tree's vegetative growth slows or stops, fruit quality will decline.

Fruiting control

(a) Crop control. When the tree is too young or the framework is as yet too weak to carry a crop of fruit then the majority of the fruit bud needs to be cut off at pruning time. This clearly depends on the ability to recognise the amount of fruit bud on the tree.

When the tree is mature enough and strong enough there is still a need for the selective removal of fruit bud when the bud density is clearly too high. In apples heavy pruning of fruit bud in an on-crop year is the first step in preventing the alternate cropping syndrome. It must be followed by adequate chemical thinning during the blossom and fruit-set period and follow-up hand or mechanical thinning.

With stone fruit it is even more important because we do not as yet have a reliable method of chemical thinning. Stone fruit trees must be heavily pruned of fruit bud when it is obvious the fruit bud density is far too high. It is commonly necessary to prune off 95% of the fruit wood.

A trap in this approach is that flowers on a heavily pruned tree are very much stronger because there are fewer flowers drawing on the tree's food and energy reserves. This results in a much higher percentage of fruit set of the remaining flowers. Follow-up hand thinning is therefore critically important.

The more the fruit tree variety is a biennial or alternate bearer the more important is flower bud pruning.

(b) Pruning must be used to assist in the **continuity of cropping.** Fruit trees like peaches, where the flower buds can only be used once and are produced along with the vegetative growth each season, must be pruned to guarantee return bloom. Basically the more you prune off a peach tree, the more it will grow back. As this regrowth contains the fruit bud, by pruning heavily this winter you have guaranteed a flowering in the season after next. The amount of flower you have this spring is largely dependent on the amount of new growth last season. This in turn may be dependent on the amount of pruning last winter.

With peaches you will leave on the tree only as much of last season's new growth as carries enough fruit bud for this season's crop. All the rest is pruned off back to a bud. The tree reacts by growing back this season all of the growth pruned off. This then provides the fruit bud for the following season's crop.

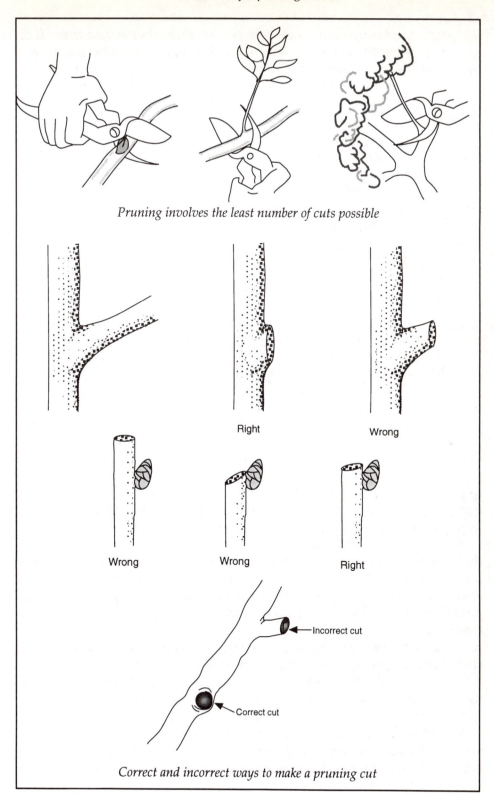

Pruning involves the least number of cuts possible

Right   Wrong

Wrong   Wrong   Right

Incorrect cut

Correct cut

Correct and incorrect ways to make a pruning cut

Grapes (and roses) grow shoots in spring, then flowers and fruit all on the shoots grown in the same season. One way of guaranteeing strong growth in this season is to prune off all shoots not necessary for the plant framework, back to a stub with one or two buds. The only shoot growth from last season which you would leave is that needed to establish a permanent framework.

With plums (especially European plums) new growth carries only vegetative buds. These develop into fruit buds on spurs during the next season. In the third season these buds produce flowers and then fruit. This means that with European plums pruning must:

- leave some one year old shoots (for fruit bud development this season and fruiting the season after)
- leave some two year old shoots because they will have flowers and fruit this year
- be heavy enough to promote new growth as this will carry fruit in three seasons. This is achieved by pruning spent and unwanted growth back to a bud.

Japanese plums and apricots fit into the same pattern except they retain fruiting spurs for more than one year.

With apples and pears fruit bud can develop on two year old growth and older, especially if that growth is at a broad angle to the vertical. With varieties which "spur" easily, very short side shoots can develop even on one year wood and every bud on this short shoot is a mixed bud. This compact bunch of one or more mixed buds is called a spur. On pome fruit, mixed buds (fruit buds) can be retained indefinitely especially if they are not excessively shaded. Often these spurs continue to develop to the stage where there are too many buds for the translocation capacity of the stem connection. Spurs should then be thinned during pruning to re-establish the appropriate balance of vigour. Many of the new cultivars of Red Delicious such as Red Chief and Oregon Spur are spur bearing varieties. Spur varieties commonly produce far more fruit bud than is needed for adequate cropping. Excess fruiting wood must be pruned off by removing the weakest shoots and those in the poorest positions. The new shoots produced by this pruning create a more even balance between young growth and fruiting wood.

Some varieties often produce fruit bud on the tips of weak or horizontal shoots and these often develop into a swollen shoot tip with several buds called a bourse. Tip bearers include Granny Smith and Rome Beauty apple, and pear varieties such as Beurre Bosc and Packhams Triumph.

Cherry trees also produce perennial fruiting spurs. These spurs are commonly concentrated near the end of each year's shoot growth. It is sometimes a little difficult to identify fruit bud on cherries. If buds are singles or in twos all buds are likely to be vegetative buds. Where there is a number of buds at each node the majority are likely to be fruit buds.

(c) A very simple concept in fruit tree management is often overlooked. A fruit tree consists of two separate but totally dependent parts. One is the root system which takes up water and nutrients from the soil, acts as a storage organ from which the whole plant can draw, and produces a share of the plant hormones.

The other half of the tree is the shoot system which carries the leaves and the flowers/fruits. Functions of the shoot system include photosynthesis using the water and nutrients supplied by the roots; transpiration metabolism, which manufactures the several thousand

Figure 4.6 *Cherry tree showing heavy fruit bud development*

chemicals plants require to operate, grow and reproduce; and the production of a different group of hormones.

When the productive capacity of the root system is in balance with the productive capacity of the shoot system then the tree will grow normally, produce to capacity and produce quality flowers and fruits. If the capacities are out of balance then several aspects of tree growth and production will suffer.

One of the reasons for pruning fruit trees is to keep the two systems in **balance.** For example if there are too many flowers for the tree to have the ability to carry as fruit, then we should remove the excess of flowers. If the tree is not producing adequate flowers for a commercial crop then pruning (and training) may be needed.

More fundamentally, the leaves can produce food from photosynthesis only in proportion to the water and nutrients supplied by the roots. At the same time the roots can only grow, searching for nutrients and water and then absorb them, in proportion to the food and energy supplied to the roots from the leaves. The two parts of the tree are totally interdependent.

Pruning can be used to help correct any imbalance. Under normal growing conditions if the tree is not producing new shoot growth this is usually a sign that there is too much shoot system already. Pruning the tree will make it grow new shoots (to replace what has been pruned off). A tree with fruit which is also growing new shoots is a tree which is in balance. The new shoot growth is very important for another reason. The most efficient part of the tree in photosynthesis is the new shoot growth. Its leaf density is higher, its average leaf size is larger and its metabolic activity rate is greater than those of any other part of the tree.

Crop management and fruit quality

(a) Fruit **size** is a very important fruit quality parameter. For most fruits, the larger the better; in some like apples and cherries only the largest fruit consistently attract viable returns. While pruning is not the only factor which determines fruit size (others include irrigation, fertilising, disease control etc.), it can contribute to maximising fruit size in two ways.

The first is to remove much of the excess fruit at the flower bud stage by appropriate winter pruning. The earlier the winter pruning the better.

The second is to maximise sunlight penetration to the fruit. What is commonly not recognised is that the fruit itself is a major photosynthesising organ. If grown in high levels of direct sunlight the fruit manufactures about half of its own building materials. This means that fruit grown in a tree with good sunlight penetration will for that reason alone be bigger.

(b) Another important measure of fruit quality is **colour**. Poorly coloured fruit attracts such low market returns it is not worth marketing. Therefore it should not be produced in the first place. Well coloured fruit will usually attract a price premium of about 50% over reasonably coloured fruit. Shopper surveys clearly indicate that consumers shop with their eyes, and colour attracts. Marketers, particularly those in the retail industry, have long recognised this fact. These marketers then set the standards for the fruit which they will buy. This determines the market standard which the growers must achieve to remain viable. This explains the constant trend to planting better coloured cultivars of all coloured fruits.

The colour development can be either physiological or genetic. Some new cultivars have genetic colour which means the fruit will achieve high colour no matter what the growing conditions are. Varieties such as these are very much in the minority.

Most modern cultivars have physiological colour. This means that the fruit has the potential to achieve good colour but will only colour well if the appropriate growing conditions prevail. By far the most important of these conditions is sunlight. Such cultivars only achieve optimum colour if the fruit is grown in near direct sunlight especially in the last four weeks of the growing season. For this reason the tree must be pruned in such a way as to allow direct sunlight onto the fruit for approximately 75% of the day. The less the direct sunlight the less will be the colour development.

In some fruits such as peaches, summer pruning is necessary as well as dormant pruning to achieve optimum colour development. The feather nature of peach primary shoot growth reinforces the need for this practice.

(c) The third parameter of fruit quality is not as obvious as the previous two. It is the **internal qualities** of the fruit and includes firmness, sweetness, flavour and aroma. These quality features are influenced by a large range of orchard management techniques. But a fundamental one is to let the fruit produce its own quality, by letting it manufacture its own materials by photosynthesis. This it can only do adequately if the fruit is grown in about 75% direct sunlight.

Recent research indicates clearly that fruits draw manufactured foods only from themselves and immediately adjacent leaves. Therefore to optimise the above four quality parameters, the fruit **and** its immediately adjacent leaves must be in about 75% direct sunlight. Correct training and pruning must be used to achieve this.

Figure 4.7 *Japanese peach orchard with the fruit growing in treated paper bags. Three different size bags are used during the growing season*

(d) The fourth parameter of fruit quality influenced by pruning is **skin quality**. Again other factors (for example frost, disease, wind, chemical burn etc.) can have an influence but pruning can assist in minimising skin damage from abrasion by training and pruning so that the fruit hangs in a space for the season and does not excessively rub against stems, buds or even leaves. The Japanese aim for ultimate perfection of skin quality by protecting the fruit for the whole season inside specially prepared bags. A price paid for this perfection is that a fruit grown in a bag cannot achieve the previous four quality parameters. Their fruit is less sweet, less flavoured and less perfumed. Large size is achieved only by thinning much more severely than is our custom.

(e) Another important aspect of adequate sunlight penetration should now be more obvious. The foods produced by **photosynthesising** leaves are either pumped to the roots, pumped into storage in the root and stem systems or used very close by. The buds are fed and maintained only by the foods produced by the leaves growing from those buds. This can only be adequate if those leaves get sufficient direct sunlight to manufacture those foods. Leaves in near constant shade cannot produce enough food to maintain the buds, so the buds die.

Buds in shaded parts of trees cannot manufacture enough food to produce fruit bud or flowers or fruit. In fact such buds cannot get enough food to stay alive. One of the major reasons for the advent of new training systems is to minimise this problem. All training systems attempt to maximise photosynthetic efficiency. This means that no part of the tree is consistently shaded throughout the season. Whether by pruning or by training, shading must be minimised.

Research into such factors has produced some valuable additional information. For example spur bearing trees are far more efficient in photosynthesising than lateral bearing trees so that they consistently grow bigger fruit. One measure of this is the number of leaves needed to feed a fruit on a spur bearer compared to a lateral bearer. Lateral bearers need about 50 leaves per fruit while spur bearers need about 15 leaves per fruit.

(f) An important area of crop management influenced by pruning is **pest and disease control**. Methods of pest and disease control are improving and developing rapidly. But fruit trees will need to be sprayed for this purpose for a long time yet. The nature of the spray material is changing as very specific pesticides are developed or materials such as insect growth regulators (IGRs) are used. Because of the economics of fruit production most spraying is done by drive-past air-blast sprayers. To obtain adequate spray coverage with the minimum amount of pesticide the tree must by open enough to allow even spray penetration.

In practice if a tree is pruned for penetration of light it is likely to be adequately pruned for spray penetration as well. When training and pruning the tree keep in mind that the spray will be coming from one direction on each side of the tree. Make sure there are adequate "windows" to ensure even coverage of the tree from the applied spray

(g) The final area of crop management is that of adequate human **access** to the tree. Many management tasks require people to be able to get into the tree. Examples include pruning, tree training, fruit thinning, pest and disease monitoring, monitoring beneficial organisms (e.g. predators), and fitting pheromone traps, lures and mating disrupters.

The culmination of all this activity is the harvesting of the crop. All fresh fruit is still harvested by hand and with the increasing emphasis on top quality there may be two or more picks per crop. The pickers need easy visual and physical access to do this job properly. This is best achieved in the earlier training and pruning programs.

Pruning in an integrated pest and disease control program

(a) A tree can be physically damaged by a number of factors including:
- crop load
- machinery damage
- storms
- hail
- rough pickers.

Where branches are broken or split, rough wounds including splits must never be allowed to remain on the tree. The branch must be cut back to undamaged areas. The point at which the cut back is made is chosen to re-establish as well as possible the original training system. Remember that the tree will attempt to grow back the branch material removed. The branch must therefore be cut back to a new leader which will best achieve this aim by directing the regrowth in the required direction.

Whether a major cut is made for purely pruning and training reasons or whether it is to restore a damaged area on the tree, such cuts should be sealed shortly after cutting. The **sealing** is to minimise the entry of pathogens to which the tree may be vulnerable. These include fungi and a range of bacteria. There is a range of sealing materials on the market and if used according to directions all give good results.

One type is a water soluble bitumen product with or without additives such as fungicides or bactericides. When the seal on a container of this product is broken the contents must be supplied with water to prevent drying out. When the material dries out it sets irreversibly into a solid form. When using the material it should be kept in a state like thick cream. If there is too much water the material is too thin to seal and tends to run off the wood. A disadvantage of this type of material is that it may crack when it dries out during the summer, breaking the seal. A second application may be necessary to maintain the seal.

Another group of sealants is related to plastic house exterior paints (with or without added pesticides). They have the advantage of being a little more flexible but commonly need two or more follow-up treatments through the season.

A third group is based on waxes. They tend to melt and run in the hot weather. Bees commonly steal the material.

(b) One major reason for repairing tree damage and sealing major cuts is to prevent invasion of pathogens which can kill or seriously damage fruit trees. There are two main groups of such pathogens.

Fungi

Some of the class of fungi called the Basidiomycetes cause common internal woodrot diseases of fruit trees. These fungi are related to the mushrooms and commonly produce "brackets" on the trees or "mushrooms" from the soil after the tree has been killed. In apples and pears a fungus called "polystictus" invades through open wounds as spores. The spores germinate and the fungus feeds on the tree internally, killing it. The outer layer of bark falls off in characteristic papery layers. If an infected branch is cut a "water soaked" area is obvious in the food storage in the wood. If the branch can be cut off well below any water soaked area and the wound sealed, there is a good chance of saving the rest of the tree. On pome and stone fruit several

Figure 4.8 *Internal wood rot fungus of fruit trees (this example is of Trametes versicolor)*

species of bracket fungi attack through open wounds and the wood of the tree becomes very soft and crumbly. Coloured brackets can be seen growing on the outside of the stem.

There is no cure for these internal wood rot fungi. Prevention is achieved by sealing all major cuts and repairing all branch damage.

Bacteria

A number of bacteria can cause severe damage and death to fruit trees. A common one in Australia is *Pseudomonas syringae* which causes a disease called bacterial canker of stone fruit trees. Cherry trees are particularly sensitive. Apricots are also very vulnerable. Other stone fruit can tolerate the bacteria if they are otherwise healthy and undamaged. A block of plum trees was observed to die from this disease after the trees were severely damaged in a hail storm and the bacteria entered the cuts caused by the hail.

Figure 4.9 *Bacterial canker of cherry trees*

This bacterium usually invades cherries through natural wounds such as leaf scars and scale leaf scars. This invasion can be minimised with an appropriate spray program. The bacteria also invade through wounds including pruning cuts. Susceptible trees should not be pruned during winter for this reason but preferably during blossom time or after fruit harvest. Major pruning cuts must be sealed to prevent bacteria entering.

(c) Many diseases of fruit trees **overwinter** on diseased parts of the tree. Examples include Leaf Curl fungus of peaches and Brown Rot blight on a range of stone fruits. Other stone fruit diseases which overwinter this way include bacterial and fungal shothole, freckle and rust. On apples, powdery mildew overwinters on blighted shoots. A major step in the control of all these diseases is to prune off all infected tissue. This is particularly important if the grower is trying to minimise the use of pesticides.

Ideally the infected prunings should be removed from the orchard and burnt. In practice if they are dropped to the ground, risk of infection will be greatly reduced and it can be reduced further if the orchard floor is treated with nitrogen (e.g. urea) to accelerate the breakdown of the prunings.

(d) Infested wood should also be removed during the pruning season to assist in the pest control program. Infestation of scale insects can best be seen while pruning. If the worst of the infestation is pruned off, removed and burnt, and the tree opened up for spray penetration and coverage, control will be much easier next season. San Jose scale of deciduous fruits, Californian red scale and white louse scale of citrus need this detailed approach for adequate control.

Shoots infested with other pests such as borers and gallwasps should also be removed at pruning time to make control easier.

In deciduous fruit trees the winter pruning season is the best time to achieve all of the above strategies. With the leaves off the tree, the framework is much more obvious and any damage, infection or infestation can be clearly observed and assessed.

Figure 4.10 *Shoot blight (brown rot) of stone fruit especially peaches and nectarines*

CHAPTER 5

The practice of pruning

The pruning procedure will depend on decisions made some time previously when the orchard was planned and designed. These decisions include:

- which fruit variety
- which rootstock
- which training system
- the maximum size the tree will grow to (determined largely by the choice of rootstock)
- tree spacing along the row
- spacing between the rows
- the support system (if any) e.g. trellising
- training aids
- pruning equipment.

No grower should commit to an orchard system unless he feels comfortable with all the requirements of that system in regard to training and pruning. The higher the density of the planting, the higher the capital investment and the higher the risk if the system is not managed successfully. Many high density plantings have failed financially for this reason.

Planting

We will assume that all appropriate site preparation and planting procedures have been followed. At planting the normal procedure is to head the tree to compensate for the root and root hair loss during nursery harvesting, transport and orchard site planting. Heading is not always necessary but it is the safest option. To decide not to head commits one to a much more precise management, especially during the first season. That entails perfect weed control, moisture management and fertilising in a suitably prepared orchard site. Heading is the only option if the nursery trees are not top quality, are damaged, or are not already the appropriate shape for the training system.

The training system then determines the correct tree management. If the orchard design includes an imposed tree shape it is helpful if the tree is oriented at planting to best use the nursery tree's natural branching. Training systems are described in the next chapter.

If the heading back can leave a clearly defined leader (or leaders) with a healthy apical meristem, the pruner has imposed control over the growth in the first season. This is important in all fruit tree varieties but is particularly needful in peaches.

Trees with naturally weak apical dominance have to be assisted in maximising the dominance. For example in spur type Red Delicious it is usually necessary to head back severely to induce regrowth and then to protect the new developing leader by either heading back competing leaders or by spreading them to reduce their apical dominance.

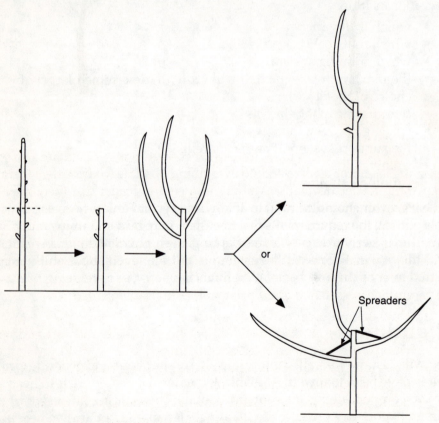

Figure 5.1 *Central leader training from a headed tree.*

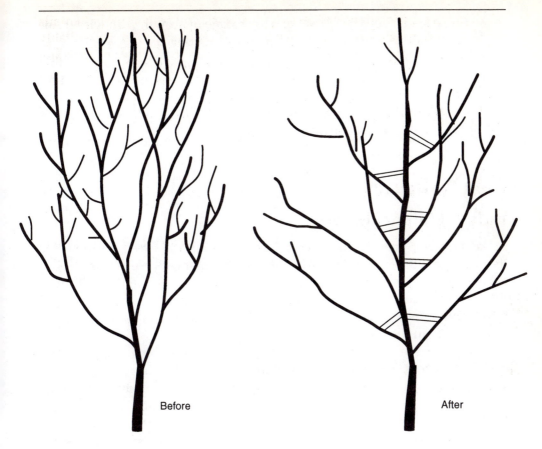

Before   After

Figure 5.2 *Wooden spreaders used to open up a tree.*

Whichever method is used, during the first growing season the aim is to select the leader or leaders and do everything possible to encourage them. If one strong leader develops naturally in a central leader tree there may be little if anything to prune or train. If however several leaders develop in what is supposed to be a central leader tree there may be some work to do. This situation is likely to arise where trees have been heavily headed at planting. With the original dominant meristem pruned off, several buds will compete to take over as the new leader. When this happens the developing tree must be carefully assessed and appropriate remedies implemented:

- Unwanted leaders can be cut off.
- They may be pinched back (when the growth is still soft and non-woody, the tip can be broken using fingers).
- They may be cut back to a horizontal pointing bud (thus weakening apical dominance in the new shoot).
- Some leaders can be spread when very young with an aid (e.g. a clothes peg) which weakens their dominance and hence their competition).

- Some leaders can be spread at a later stage of growth with wire trainers, small weights or some other spreading device.

In all of the above examples we are managing apical dominance and this may or may not include pruning. What it does require is an understanding of how the tree grows and an understanding of apical dominance. This still applies in trees which are 10 or 30 or 50 years old. That is why we discussed apical dominance before we discuss pruning.

General rules

Rule 1: Major cuts first

Make the major pruning cuts first. Take a good look at the tree, decide what its structural problems are (if any) and which major cuts are required. Forget any detailed pruning until you have done this. For the first two or three growing seasons after planting there is a tendency to let the tree do its own thing. Only limbs which are clearly contrary to the training system are removed early in the tree's life.

Leaving the leaders unheaded and doing their own thing will grow the tree's framework in the quickest possible time. Other pruning normally consists of thinning out side shoots (potential fruiting arms) and directing them to a sidewards growing leader. This prevents them competing with the tree's leaders and also develops strong fruiting arms in the shortest time.

With trees on dwarfing rootstocks (e.g. M9 or M27 in apples) or specific shape training, there may be a need to support the leader to grow in a particular orientation. This support may be in the form of posts, suspended rods, wires, trainers or other aids.

The major cuts should be made in such a way as to control the regrowth. This means that in most (but not all) cases the cut will be made at a point where a bud, spur or shoot will develop the new leader. The growth of the new leader will be in a direction you have chosen. Where a new leader is not required because the tree already has enough, then cut to a collar so that the regrowth will develop in the immediately adjacent branch.

Rule 2: Dwarfing effect

The way to grow the framework most quickly is not to head and to leave the leaders alone because pruning is a dwarfing process. The more the tree is pruned, the longer it takes to develop its framework and to fill the allocated orchard space. The training system which uses this concept to the extreme is the Spanish bush system for cherries.

This rule does not imply that no pruning should be done. As the main framework develops along with the fruiting arms there will be situations where there are too many branches in a space or fruiting arms are developing too closely together. Pruning must be done to remedy such situations and if it is done early and properly, it will have no dwarfing effect at all.

The main framework (with necessary adjustments) must be left to do its job and grow the tree size determined by the rootstock. When the main framework grows outside its allocated orchard space then it too must be pruned.

Similarly when the developing framework produces excessive shading then this also must be remedied. By examining the tree closely it should be obvious in such situations which major cuts may need to be made to eliminate major shading.

Rule 3: Prune to invigorate

Pruning invigorates the tree but at the expense of the original tree size and the number of growing points. Remember that the best fruit is grown on a tree which is growing. Pruning may be necessary to invigorate the tree if the growth has been declining. This is especially important with peaches where fruit bud is produced only on this season's shoot growth. No shoot growth this season means no peaches next season. This is why peach trees over ten years of age appear scalped when pruned appropriately for the tree's condition.

One of the traps with this invigoration is that excess shading may be created by the heavy pruning. The answer is to prune back to newly elected leaders which are in the best position and point in the best direction for the regrowth. Do not make excess work for yourself next year by pruning badly this year.

Rule 4: Automatic reaction

Fruit trees react automatically — not intelligently —to pruning and training. If you understand the nature of this automatic response then pruning is very obvious. For example, trees react to being pruned by doing their utmost to grow back what has been cut off. You can use this reaction to prune off unwanted growth and direct the regrowth to where it will be useful. All you need to do is to cut back to an appropriate new leader pointing in the best direction. This new leader may be a weak, nearly horizontal shoot. But if it is best placed and is pointing in the best direction then it is the best new leader.

You lose this advantage if you head this new leader. Remember the control you have over the growth of the tree is the apical dominance of the leader no matter how weak it is. Use the apical dominance you have. Heading only produces a number of competing new leaders which cause congestion and shading.

Rule 5: Anticipate problems

Pruning may increase tree problems. Where a major limb is cut out and no provision made for where the regrowth must go, excess shading will be the result. So there are two closely related problems to solve. A decision must be made when a major limb is to be removed by pruning. The tree will try to grow back the removed limb next season. So in cutting out the limb planning must be implemented for the replacement growth. Cut to appropriate new leaders.

Rule 6: Pruning delays fruiting

Pruning delays fruit production. Because the reaction of a tree to major pruning is to attempt to grow back the pruned material, fruit bud production is delayed while new vegetative growth is produced.

When trees are allowed to grow as described previously, the weaker side shoots will develop fruit bud very easily on shoots that have very weak apical dominance. Pruning these shoots to strengthen them delays fruit bud production.

The only trees which produce some fruiting response to pruning are peaches, where fruit bud is found only on one year wood, and plums, where fruit bud is found on two year wood.

Rule 7: Reaction is localised

Pruning has dominantly localised effects. Major limbs on a tree each react as if they were the whole tree where pruning is performed. The tree will attempt to regenerate the growth at the point where it was cut off.

Rule 8: Disease enters through large cuts

Large pruning cuts increase the risk of disease. As discussed in the previous chapter, a number of destructive pathogens find entry into fruit trees via large wounds such as large pruning cuts. This fact is not an argument for making no major pruning cuts. Rather it must be taken into account when planning a pruning program which may include major cuts.

For example, cherry trees are known to be very vulnerable to attack from the bacterium *Pseudomonas syringae* which causes the disease known as bacterial canker often leading to tree death. This factor is included in good planning for the pruning of a block of cherry trees. Knowing that the bacterium is very active in wet weather and in cold temperatures suggests that cherries should be pruned when such weather conditions are unlikely. So pruning of cherry trees is often recommended for the hot dry weather immediately after harvest.

This is fine as it stands. But it is equally important to minimise infection risk by pruning in such a way that the normal healing processes of the tree can occur in the shortest possible time. If the tree is under drought stress its healing processes are severely reduced. The wound will not callus or heal quickly. This factor suggests that pruning of cherries be done when vegetative growth is at its maximum. That time is in the spring, from late blossom time to petal fall time.

Cherry trees heavily pruned in mid-summer have often died not from bacterial canker but from the entry of the internal wood rot bracket fungi. The fungal spores are blown around by the wind even in hot dry weather. An open wound on the tree is an invitation to a free feed for the fungus.

All fruit trees are vulnerable to internal wood rot fungi. In some fruits, this type of damage is very common. For example, apples are extremely sensitive to the wood rot fungus called polystictus, killing entire orchards which have been badly pruned (often over a period of years).

The lesson from all of this is to choose the timing and implementation of severe pruning so that such risks are minimised. When large cuts are made it is strongly recommended that the wound be sealed with an appropriate material with follow-up treatment as required. This recommendation also applies when trees are cut down for "top-working", that is, to change the variety of the top of the tree by grafting. Many grafts which would otherwise be successful have failed because of entry into the wounds of pathogens such as bracket fungi.

Rule 9: Some trees need summer pruning

Summer pruning is often required especially on fruit varieties which produce a great deal of vegetative growth during the growing season. While summer pruning is used for a variety of tree management reasons, the most common is to maximise fruit colour.

As discussed previously, the fruit itself is a major photosynthesising organ. It is often the main source of the building materials which produce fruit quality characteristics such as size, colour, sugar levels, flavours and aromas. With the intense pressure in the market place and especially the buying criteria of the major chain stores, the production of highly coloured fruit is a must.

Therefore in fruits such as peaches where feathered summer growth is the norm it is often necessary to prune in summer at the appropriate time to get direct light to the fruit so that maximum colour can be achieved. The critical time for fruit colouring in peaches is about the last four weeks prior to harvest. Therefore the shoots causing the excessive shading should be pruned off at the beginning of this period. If the summer pruning is also necessary for other fruit quality features (e.g. size, sugar levels etc.) the summer pruning may need to be somewhat earlier.

The timing is critical especially in low chill varieties where the fruit growing season is so short. If the summer pruning is done too early the tree will regrow the pruned material and shading will quickly be re-established. If the pruning is done too late then the fruit has not been given the opportunity to produce the optimum colour.

There is no one point in time which suits all fruits or all cultivars. The optimum time is usually best defined by identifying the particular point in the tree‚s growth curve. In peaches it is that point on the growth curve where the fruit is rapidly increasing in weight. This is the bottom of the second "S" in the double sigmoidal growth curve. This point will be at a different date for each cultivar but will be at a similar time for cultivars maturing close together.

Apples have a single sigmoidal growth curve and the increase in fruit size and weight is far more gradual than in stone fruit. There is also a big variation between cultivars as to when colour development begins. In general however a red variety of apple should have about 70% incident light for at least four to six weeks prior to harvest date.

The growth curves in the diagram are created by mapping the increase in the fresh weight of the fruit against the days from full bloom. There are no numbers on the graph because it summarises general tendencies. Every cultivar has a different graph.

At full bloom, fertilisation occurs and the cells in the young fruit undergo very rapid cell division. All the cells a fruit will ever have are produced in the three weeks or so after full bloom.

The fruit swells in size quickly during this period and there is a great deal of hormone activity. In peaches (and most stone fruit), after the three weeks fruit growth stops and nothing appears to be happening. Over the next period the pit or stone in the fruit begins its development. At the same time the peach tree enters its phase of maximum vegetative growth. After pit hardening finishes, the fruit will start to increase in size. This normally starts about four weeks prior to harvest. For peaches it is at this time that summer pruning should occur.

There is no equivalent of pit hardening in apples. When cell division has finished, cell expansion then begins, slowly at first but very gradually accelerating. Summer pruning of apples should occur about mid-way up the steep slope of the growth curve. This is normally about four to six weeks prior to harvest.

There is one other factor to keep in mind when planning pruning for colour development of fruit. The fruit of many cultivars of a variety of fruits can burn if they grow in too much direct sunlight and especially if the growing season is extremely hot and dry. Apples such as Granny Smith and Fuji sunburn very easily. Stone fruit and especially plums and nectarines can sunburn. Sunburnt fruit is unmarketable. If that were not enough the bark, especially on young trees, can sunburn very easily. The sunburnt bark

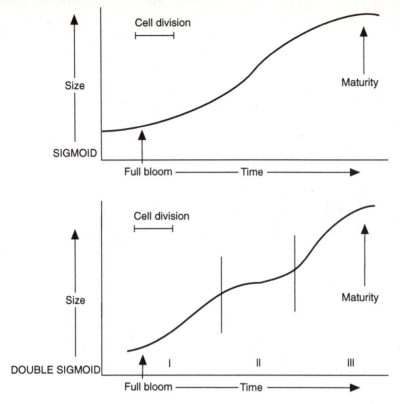

Figure 5.3 *Stages in the growth of fruit*

splits and peels from the wood allowing another point of entry for pathogens. Trees can die from severe sunburn. In training systems where branches are trained into a fairly flat position this must be kept in mind when pruning. Enough shading shoot growth must be retained to minimise sunburn damage.

To summarise these apparently conflicting requirements, the aim is to grow the majority of the fruit in about 70% to 80% sunlight for quality and colour. In all training systems some of the fruit will not be in the optimum position. To grow the fruit in 90% to 100% full sunlight risks fruit sunburn damage, especially in very hot growing seasons and especially on the more susceptible cultivars. Pruning too much, so that branches are constantly in 90% to 100% direct sunlight is likely to cause tree damage from sunburn in hot seasons.

Rule 10: Allow for the growth habit

Always take the tree's natural growth habit into account when pruning. For example extreme spur type apple trees require consistent pruning and training to achieve acceptable vegetative growth. With lateral bearers the

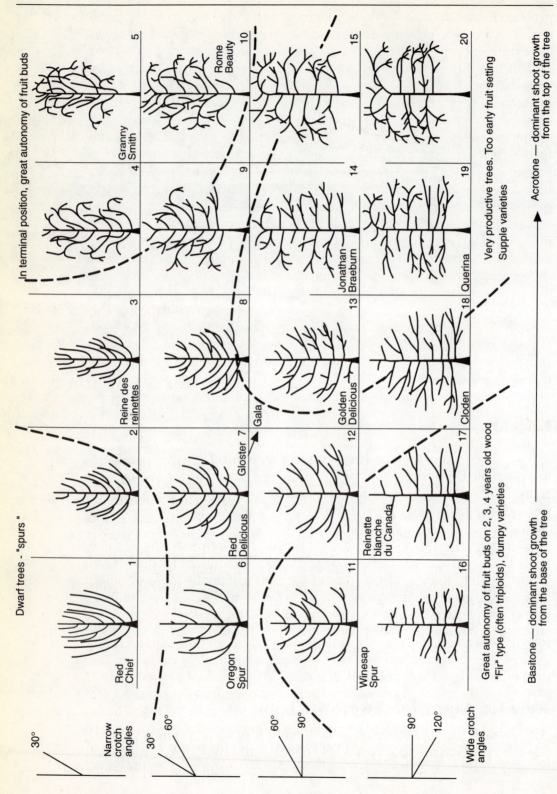

Figure 5.4 *Lespinasse's classification of apple tree types by growth habit*

problem is often too much vegetative growth which is exaggerated by pruning. Such trees need training not pruning. But various cultivars of apples have different growth habits.

This diagram summarises the range of apple tree growing habits as interpreted by M. Lespinasse of the INRA Research Station in France. Along the bottom axis the balance of strong growth ranges from basitonic (strongest vegetative growth from the bottom of the tree) on the left, to acrotonic (strongest vegetative growth from the top of the tree). On the vertical axis the vegetative growth reflects extremely strong competition for leaders (resulting in narrow branch angles) at the top to wide crotch angles (hence little leader competition) at the bottom. Every commercial apple variety grown today fits in to the combination of these two factors.

Clearly the specific growth habit of each apple tree type must be taken into account when pruning and training.

Peach trees grow in a way very different from apples. There is the basic difference that peach trees fruit only on one year old wood where apple trees can fruit on wood of any age. Another basic difference requires a different pruning and training style. Strong peach growth is always feathered. This means that when a new shoot starts growing in the spring the initial growth is new growth (primary growth) adding a new length of shoot. But as this shoot grows some of the buds produced are not inhibited by the apical meristem as we discussed earlier. This suggests that the balance of hormones in the peach shoot is different from that in most apically dominant trees.

The net result is that the main shoot increases in length and during this period of growth, some of the buds produce side shoots. The result is what has been called feathering. A consequence of this feathering is that peach trees produce very large areas of leaf very quickly. This leads in turn to the shading and the need for summer pruning as discussed earlier. But it also means that peach trees have the potential to produce a strong tree framework of suitable size very quickly. Because many of the feather side shoots especially on vigorously growing trees will carry some flower bud, there is very significant potential for early cropping. The feathering also gives many more options for the directing of the growth in the pruning season.

There are four basic approaches to pruning strongly feathered growth on peach trees:

1. Leave the dominant apical meristem alone (that is, do not head the tree) but cut back short most of the side shoots. This will minimise shading developing in the tree and also help reduce excessive early cropping. This method will grow the framework of the tree in the quickest possible time. It also means the tree will grow too tall very quickly. But as long as the apical meristem is left to dominate, you have major control over the tree. The excess height can be cut back to a new lower leader once the tree framework has been established. This concept applies whether the peach tree is grown in a vase, central leader, palmette, Tatura trellis or any other configuration.

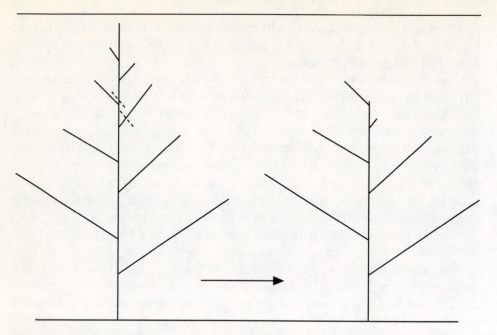

Figure 5.5 *Reduction of height of tree while maintaining apical dominance control*

2. Prune the top of the main shoot back to a suitable side shoot which then becomes the leader. Other excess side shoots are removed to prevent excess shading. This method may help to spread the tree earlier which will reduce shade. It will also mean the tree will not be as tall.

3. Cut most of the side shoots back to one or two nodes. This will reduce excess early cropping. But this method suffers from the same problem that all pruning based on heading back shares. Whenever any shoot is headed the single shoot will be replaced by several shoots as the buds left compete to become leader. It does not matter whether the headed back shoot is the main vertical leader in the tree or a near horizontal, weak side shoot: the net result will be the same. The only difference will be in the vigour of the new shoots.

4. The standard method has been to head the tree heavily at planting and then head back each season,s growth. This method has the same problem as 3, but more so. A constantly headed tree does not know which way it should be growing so it grows every way. This method violates the basic concept of pruning which is to help the tree to grow best by choosing leaders (and therefore, apical dominance) and leaving them to it.

There is another problem it creates in peach trees. At every node on the new growth there will be a vegetative bud and if the growth is strong, there will be one or two flower buds. After heading back the vegetative buds are no longer inhibited and will compete to become the new leader. With heavy heading, pruning flower buds near the pruning cut will revert to shoot buds. This further increases the tendency to excess shading.

Figure 5.6 *A new leader selected in a peach tree*

So while heavy pruning of peach trees is often needed for crop control it is better to cut a main shoot back to one new leader than to head each side shoot back.

Rule 11: Peaches and cherries

Peaches and cherries have a distinctive growing behaviour which does not quite fit into the model of tree growth we have discussed. It only becomes useful when heading back has been largely deleted from our approach to pruning. The emphasis so far has been on the (oversimplified but adequate)

Figure 5.7 *Three shoot development from the apex of a non-headed cherry limb. The terminal bud and two sub-terminal buds have each grown vegetatively*

concept of apical dominance. It relies on the apical meristem dominating those buds below it and in general inhibiting them from developing side shoots.

If you look carefully at the tip of a peach or cherry shoot in the winter you will see the normal apical bud and below it the nodes with their buds. But at the tip you will see usually two but often three or four buds almost beside the apical bud. In the spring these buds show no sign of being inhibited by the apical bud and readily grow side shoots. These side shoots are at a slightly narrower angle than the other side shoots but are also a little stronger.

With selection these buds can lead to the development of strong fruiting arms and even potential new leaders if the tree height has to be reduced in the future. These shoots are especially usefull on trellised trees where there is a need to train the tree to spread on the trellis. But even on free standing trees these shoots offer more options when deciding on the framework of the tree.

Rule 12: Pruning influences nutrient distribution

Pruning influences the movement of nutrients in the tree. The best example is calcium. Calcium is a relatively immobile nutrient; trees have some difficulty in taking it up from the soil even when soil conditions are near optimum and calcium also does not move very easily within the tree.

A major role of calcium in the tree is the development of the strength of the cell walls in both young growth and fruit growth. Young vigorous shoot growth tends to have priority over the calcium supply. This is a problem with those cultivars which need high levels of calcium for fruit quality, especially crispness, flesh quality and shelf life.

When such trees are pruned the available calcium is drawn to the vegetative regrowth and the fruits will suffer deficiencies. For such varieties pruning and nutrient management are closely linked.

Rule 13: Limit disruption

Pruning and tree training consist of manipulating the natural growing nature of the tree. The less massive and disruptive the adjustment of the tree can be, the more successful will be the procedure. By assisting apical dominance and managing the tree the best results will be achieved.

So training is as important as pruning and in many cases is the preferred approach. Spreading limbs which are not wanted as leaders is better than drastic pruning. Training systems commonly place branches into a preferred position. Branches are tied to trellising, or training aids are used to control the branch angles. In very intensive orchards, fruiting arms are commonly tied up to support posts to stop them dropping below horizontal under the weight of a crop. The reasons for all of these strategies should be clear in the chapter on apical dominance.

Plant growth regulators can be used as an aid to tree training as well but this technology must be thoroughly understood before attempting to use it.

Another strategy which can be used is notching. This consists of making a small cut in the stem near a bud to trigger a change in development. Usually a simple cut into the bark to the cambium layer is all that is required. In other approaches the cut is made by a hacksaw blade which creates a wider wound.

If the cut is made above the bud, the downward flow of auxin is temporarily halted. The inhibition on the bud is removed for a short period. This is sufficient time to trigger the bud into becoming vegetative and a shoot will develop. If the cut is made below the bud it acts like a dam to bank up the auxin and the food from the leaves. The result this time is to trigger the bud into developing flowers.

The wounds heal very quickly and normal tree development continues. But the slight short period changes to auxin flow are sufficient to trigger the developmental responses.

The plant growth regulator Cytolin can also be used to initiate branching where it is required, without the need for pruning.

Rule 14: Prune individually

There is no one correct way to prune or to train a fruit tree. If you asked twenty different experienced pruners to prune the one tree, no two approaches will be identical. But each would have achieved the results discussed in the previous chapter.

The way to prune a tree is to keep in mind the reasons for pruning, the selected training system and the type of tree. If the tree needs pruning then you prune it. If the tree does not need pruning then leave it alone.

The best looking tree is the one that consistently produces a good crop of quality fruit. If each tree is a different shape from its neighbours because you have let it grow its own way within reason, then so what? Orchards no longer consist of trees pruned so that they are identical in shape. Trees are pruned to be efficient producers of quality fruit.

Rule 15: Don't try to solve all problems at once

Not all of a tree's pruning or training problems can be solved in any one pruning season. This is especially so if the tree has been neglected or badly pruned in the past. Remember that the tree will attempt to regrow next season all wood pruned off. So heavy pruning will mean vigorous regrowth which may or may not develop where you would want it.

The other problem is that the more regrowth there is, the more difficult it will be to maintain cropping. In addition the heavy regrowth will tend to produce shading. So trees pruned very heavily will have problems cropping consistently and the crop achieved is likely to be of poorer quality.

The correct approach is to try and solve the major problems first. This usually means thinning out excess branches. Choose the branches to remove, attempt to direct the regrowth which will result and monitor the result during the growing season and then in the next pruning season.

If there is no problem with excess branches but there are too many fruiting arms and they are too close, then the first step is to redesign the fruiting canopy by retaining the best fruiting arms and removing the worst (as far as position or shading is concerned).

The major problem may be excess height of the leaders. If this is the case then reduce the height by cutting back to a new leader at the required height. Make sure this new leader does not have competition as leader. If this is not done the regrowth at the top of the tree will be too bushy and will create shade.

CHAPTER 6

Training systems

There is a very large number of fruit tree training systems used around the world. Each one reflects the nature of the fruit growing industry of the area (including cost of orchard land, cost of labour, growing conditions and the returns for the fruit) and a continuing search for the ultimate method.

The general trends in orchard design include:

- increased production per hectare
- much smaller trees
- much higher tree densities
- lower costs of production
- more efficient use of highly skilled labour
- much earlier production (often the aim is a commercial crop two seasons after planting)
- fewer and larger orchards
- higher quality fruit to meet demand
- higher visual quality (bigger, better coloured fruit).
- better eating quality and better varieties (an increasing trend in very recent years).

Several factors complicate this scenario:

- Fruit markets are international.
- Increased competition comes from other convenience foods as well as new fruits and overseas suppliers.
- Consumers expect higher quality fruit.
- Grower returns have fallen sharply in real terms (i.e. in dollars per kilo).

The grower's share of the retail dollar has fallen sharply. Thirty years ago the grower share of the retail dollar for fruit and vegetables averaged about fifty cents; today few achieve greater than ten cents.

Those growers committed to the industry are becoming progressively more efficient in every aspect of fruit production. Orchard design is one of the major areas.

Because of the low returns and low margins, a grower must obtain returns from his orchard in the quickest possible time. He must get high levels of production quickly without any sacrifice of fruit quality. The average fruit quality of twenty years ago would bankrupt an orchardist today. So while early production and high production is critical, high fruit quality is even more critical.

The average orchard fifty years ago had about 80 trees to the hectare, grew a tree for fifteen years before any fruit and most of the fruit was from 2m to 10m from the ground. Today a typical tree density is from 1000 to 4000 trees/ha, with a few up to 40,000 trees/ha. Fruit is produced from year two, with heavy cropping from year four or five. Most if not all the orchard work is done from the ground because tree height is about two metres (Table 6.1).

The modern orchard is very expensive to set up and early returns are critical. Nursery trees particularly patented varieties in patented stocks are very expensive in Australia. But with the higher tree densities a great deal more skill is needed with every aspect of tree management (tree size control, pruning, training, fertilising, irrigation, pest and disease control, harvesting).

A range of tree management systems has been trialled around the world in the race to get an edge. All these systems are only as good as the skill of the manager. There have been many major economic failures even of the most modern orchards. It is critical that a grower should adopt a system only if there is a depth of understanding of the requirements of the system. The modern orchard is very unforgiving of mistakes (Table 6.2).

Clearly the trend to high density orchards has meant that trees are now much smaller. The old types of fruit trees cannot be planted closer together to produce a modern orchard. If that were done the orchard would be all wood. The only practical way to control the size of the tree is to use a size controlling rootstock. The most advanced fruit industry in this regard is the

Table 6.1 *Average apple yields in the Netherlands(tonnes/ha)*

Tree density (trees/ha)	2000	3000	4000	Difference between 2000 and 4000 (%)
Year 2	5.0	8.3	11.6	132
Year 3	16.5	23.2	30.0	81
Year 4	24.5	31.8	39.1	60
Year 5	31.4	37.7	44.1	40
Year 6	32.7	37.0	41.3	26
Year 7	35.7	39.5	43.3	22

Table 6.2 *Orchard management systems*

Very intense density and input
Over 2000 trees per hectare

System:	Trellis, hedgerows, double row, spindle bush.
Management:	Skilled management required. Irrigation and weed control essential from the first year. Summer pruning and training essential.
Cultivars:	Rootstock and cultivar choice are critical for size control.
Support:	A support system is essential.
Comments:	Requires total commitment from the grower, with a complete understanding of tree growth and nutrition.

High density and input
Over 1500 trees per hectare

Systemn:	Central leader pyramid shaped trees.
Management:	Skilled management required. Irrigation and weed control essential from the first year. Intense training.
Cultivars:	Rootstock choice is essential to size control.
Support:	A support system is essential.
Comments:	Correctly managed, this system can give high yields early in the life of the orchard.

Medium/high density and input
Approximately 1000 trees per hectare

System:	Central leader training is essential.
Management:	Careful supervision and skilled tree management are required.
Cultivars:	Consideration should be given to the cultivar/rootstock combination.
Support:	Depending on the rootstock, a single wire support system may be required.
Comments:	Possibly a viable stepping stone for growers who do not feel their skills are developed enough to proceed to the high density system.

Medium density and input
Approximately 500 trees per hectare

System:	Average training system.
Management:	Moderate work time.
Cultivars:	Less importance is placed on selection.
Support:	May not be required, depending on rootstock
Comments:	Less skills, commitment and investment are required.

Low density and input
Approxiately 250 trees per hectare.

System:	Least input of any system required.
Comments:	Not an economic system due to low income returns, especially in the early years of orchard life.

apple industry where specific size control rootstocks have been in use for decades. Further research is continuing at a very rapid pace. Rootstocks will be considered in the next chapter.

It is possible to impose some size control over trees by using training systems or by the use of plant growth regulators (such as paclobutrazol). This chapter examines some of those training strategies. Plant growth regulators are a poor substitute for the other methods discussed and should be used only if there is a thorough knowledge of this field.

The fruiting variety has some influence over tree size and again this knowledge is well developed in the apple industry. Table 6.3 gives an indication of the influence various apple cultivars have over tree size. The rootstock however is the major influence. So the grower must choose very carefully the stock/variety combination which is to be fitted into the orchard design. The combination may also have been chosen to suit a particular training system. Having done all of this the grower then has to guide the trees in the chosen training system by using his knowledge of apical dominance, plant hormone behaviour, tree growing habit and the requirements for producing crops of high quality fruit.

The training systems discussed include:

- the vase tree
- the Spanish bush (for cherries)
- central leader types

    - Central leader
    - Vertical axis/the axe
    - Spindle bush
    - Hytec

- multi-row bed systems (using central leader trained trees)
- the palmette
- oblique – Bouche-Thomas
- trellis systems

    - Vertical
    - Tatura
    - MIA
    - Yanco cantilever

- canopy trellis systems

    - Lincoln
    - Solen
    - Ebro-espalier
    - Japanese canopy

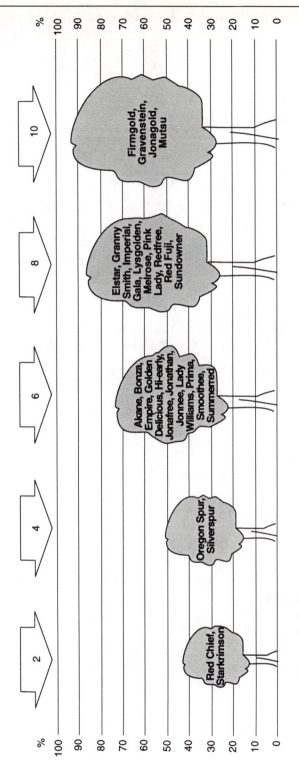

Table 6.3 *Cultivar vigour*

Figure 6.1 *Vase apple tree on seedling rootstock (height about 6m). This tree has been very well pruned for such a tree but would have taken two man-hours to prune.*

The vase

This system is the cheapest and simplest to establish. It has the least skill requirement of the operator and is tolerant of errors. The orchards of the world have been dominantly of this style and some orchards are still trained this way. The vase is an efficient way of growing fruit tree wood but in general is not an efficient way of growing fruit.

Tree densities in the past ranged from about 80 trees/ha to about 250 trees/ha. In some newer plantings tree densities approach 500 trees/ha.

Successful pruning and training of the vase tree required little skill as most pruning was done on a template basis (e.g. head every leader to about one half or cut each side shoot to three buds). All trees were originally propagated on very vigorous rootstocks such as seedling but most recent plantings have been on semi-dwarfing stocks.

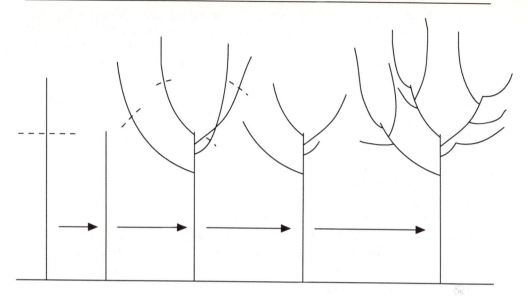

Figure 6.2 *Evolution of a vase tree*

This development occurred as it became more obvious that under most conditions fruit grown three metres or more from the ground was too expensive to grow, to thin and to harvest. An exception to these conditions is the fruit industry in the USA where growers still have access to very cheap and readily available thinners and harvesters called Mexicans. In the other developed countries, orchard labour was much more expensive and accelerated the move to more cost effective orchards.

Figure 6.3 *Vase apple tree on Northern Spy (semi-dwarfing). The tree is about 3.5m high. It now lacks vegetative growth so it should be pruned back to establish new growth.*

The tree from the nursery is usually a simple whip-stick and is headed to about 600 to 1000 mm at planting. In the spring the top few uninhibited buds develop shoots which compete as leaders and therefore grow with very narrow branch angles. This multiple leader development is strengthened by the vigorous stocks being used and these leaders develop into the main branches of the tree. In the next winter each new leader is headed again, producing several new leaders on each branch. Usually the top leader on each branch is headed less than the others and so it becomes more dominant because of the attrition of the others. One argument used for the vase system was that this heading back was needed to "stiffen" the tree in preparation for carrying the fruit crop.

This process continued until the tree came into commercial bearing. This would be about year four for peaches, year eight for semi-dwarfing apples and year fifteen for apples and cherries on seedling stock. From this point on pruning consisted mainly of pruning off the excess growth produced by last year's pruning. Every time a shoot was headed several shoots were developed which needed pruning the next year. Far too much of the tree's resources were committed to growing more tree rather than more fruit.

There should be several clear results from this type of pruning :

- The tree shape will be narrow at ground level and broader at every step of height up the tree. This leads to dominant shading of the lower part of the tree all day with obvious costs in fruit quality. With the lower buds being shaded there is a major tendency for these to die from starvation.
- Fruiting is delayed while the heading process continues. This and the tree shape mean that the best fruit is grown at the top of the tree. All management of the fruit has to be done from ladders. With expensive labour, each step up the ladder means more costs of production of the fruit.
- The production of quality fruit still requires detailed pruning at the top of the tree for adequate light penetration, also from the top of a ladder.
- Effective pest and disease control depends on high pressure or high wind power spray units plus ongoing detailed pruning for spray access windows into the tree.

Some stone fruit orchards are still trained this way with adaptations to cope with the shading problems. This includes early spreading of the main branches to produce a mushroom shaped tree. This has commonly led to sunburn of the spread branches especially in very hot seasons.

The Spanish bush

This is a system of training cherry trees developed in the cherry growing areas of Spain. The aim is to produce a cherry tree which can be largely harvested from the ground. The method does work under the specific growing conditions in Spain but there is major doubt as to whether it would work under Australian conditions.

As discussed earlier, pruning is a dwarfing process and the Spanish bush system depends on repeated heading back to control tree size. In Spain the trees are propagated on a very precocious rootstock (Mahaleb) San Lucia 64 and the successful orchards are planted in very coarse alluvial gravel soils. This in combination with the hot dry climate commonly puts trees under some water stress.

The newly planted trees are headed at about 200 mm, very close to the ground. This heading cut is usually made just after bud break, a technique which produces wide branch angles in the new leaders. After the new leaders have reached about 600 mm to 700 mm, about 250 mm of the leader is cut off. This process is repeated when the regrowth leader has again reached about 600 mm. The original three to four leaders (from the heading back at budbreak) develop about eight to ten "secondaries" so that at the end of the first growing season there are about twelve to fourteen branches. The trees are stunted by two episodes of severe pruning and by preventing apical dominance from achieving its potential.

In year two dormant pruning removes any dominant shoot and spreads the regrowth over the remaining number. Crossing branches are also removed at this stage. The primary (or leader) shoots are headed back by half in February and thinned if necessary. The lesser shoots are not pruned if at all possible but any strong shoots are headed back.

In year three the branches are thinned to maintain an open centre to obtain light penetration and again dominant shoots may be headed back. The trees are normally into production by year three or four and from this time pruning consists of thinning limbs where crowding occurs and training may include tying down of strong competing shoots.

The system depends on:

- a dwarfing and precocious rootstock
- persistent heading of the leaders
- a deficient irrigation regime
- a poor soil especially if the stock is not particularly dwarfing
- a climate which allows this pruning regime with minimum risk from bacterial canker
- the use of Cytolin to develop side shoots if the stock is too strong.

The potential for this system in Australia will depend on the availability of reliable dwarfing rootstocks under Australian conditions similar to the Giessen 148-8. As well the bacterial canker risk is likely to be too high for this severe pruning regime in present Australian cherry localities.

The central leader

The natural way for an apically dominant tree to grow is the central leader (pine tree shape). This also happens to be the most efficient tree shape in both interception of sunlight and management of fruit crops. It is far easier to train a tree for light penetration and for optimum light interception with central leader. Less feeder leaves are needed per piece of fruit in central leader. More fruit is produced per unit stem diameter in central leader trees than in any other tree design.

If this is true then why are all fruit trees not grown as central leaders ? There are several reasons. For example:

- If the rootstock is too strong you will grow telegraph poles.
- Fruits very sensitive to sunburn may get too much direct light.
- The tree trainer needs a lot of understanding of how trees grow so as to minimise problems and solve them when they occur.

Figure 6.4 *Central leader apple tree on a semi-dwarfing rootstock with a height of about 2m*

- More mistakes are made in central leader training than in any other method.
- Mistakes made when heading at planting and in the first few months of growth are often difficult to remedy.
- Vase trees are far easier to prune.
- Some trees (Nashi for example) have wood that is too brittle to support the crop weight and trellised systems are better.

Central leader trees on dwarfing rootstocks need tree support because of weak anchorage, so posts or trellising may be required. The leader of such trees also may need support because the stock has so weakened the apical dominance that vertical shoot growth cannot carry the weight. This is a common problem in the vertical axis system and thin poles are hung from support wires for the sole purpose of supporting the leader.

Central leader trees on semi-dwarfing rootstocks usually do not need any support. The problem however is the boundary between those trees which need support and those that do not. The boundary is not clear cut because factors other than the rootstock influence the strength of the tree's growth. These factors include soil depth and fertility, environmental

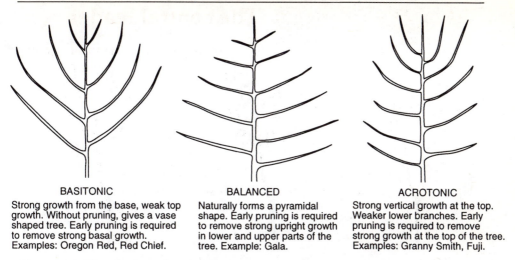

BASITONIC	BALANCED	ACROTONIC
Strong growth from the base, weak top growth. Without pruning, gives a vase shaped tree. Early pruning is required to remove strong basal growth. Examples: Oregon Red, Red Chief.	Naturally forms a pyramidal shape. Early pruning is required to remove strong upright growth in lower and upper parts of the tree. Example: Gala.	Strong vertical growth at the top. Weaker lower branches. Early pruning is required to remove strong growth at the top of the tree. Examples: Granny Smith, Fuji.

Figure 6.5 *Three basic apple tree forms*

conditions, the virus status of the tree (the stock or the scion), water and nutrient status and pest/disease status.

The art of training central leader trees is the accurate implementation of the concepts of apical dominance. The implementation of this training starts with the nursery tree. Growers should order from the nursery the scion/stock combination which suits their area. The trees should also be ordered as feathered whips. In European orchards this step is considered critical.

When the tree is planted the first steps are taken in the development of the central leader by pruning the nursery tree according to how closely it fits the ideal in diameter, length, whether the tree is feathered, at what height the feathering starts and at what height it finishes. The correct pruning of the tree according to its features is critical. We can summarise the pruning using the following examples.

(a) The nursery tree is about 1.5 m to 2 m tall from the bud union and feathered with a whorl of wide angled shoots in the range of 0.6 m to 1.0 m (Figure 6.6). The main shoot is headed about 15 mm to 20 mm above the highest retained side shoot. The selected side shoots are cut to about 10 cm to a downward pointed bud (this will help retain the broad branch angle). Other shoots including broken ones are cut back with a bench cut (a short horizontal cut).

When growth begins in the spring the top three of four buds produce vertical shoots competing to become the leader (this should be obvious from our discussions about apical dominance). The appropriate one of these is selected as the leader and the others when about 80 mm to 100 mm long are cut or pinched back to a basal bud. These will shoot again but are now inhibited by the leader and so will shoot at a broad angle. Apical dominance is now controlling the tree.

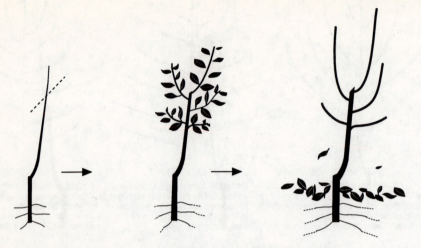

Figure 6.6 *The development of a "conventional" two year old nursery tree*

Other side shoots will be removed or pinched back depending on whether they are suitable as side shoots. These shoots should be spaced a minimum of 60 cm apart and preferably wider. This will allow for adequate sunlight penetration as the side shoots develop into fruiting arms.

The process continues for the next few growing seasons while apical dominance is maintained. Only the leader will grow vertically; all other shoots will be at a broad angle. Pruning is restricted to controlling errors in the tree's development as they occur during growth. It is possible that little pruning is needed for several years.

If competitors for the leader develop then they should be removed. If the leader is weak (e.g. dwarfing rootstock) then support it. Spreaders and other training aids may be necessary to maintain broad branch angles. If wire spreaders are used (especially on stone fruit) leave them on the tree for no more that eight weeks (October to November).

When the tree framework is adequately developed but the leader is now too tall, choose a new leader at an appropriate height and cut the original leader back to it. When side shoots become too long shorten them back to a new suitable leader. If side shoots

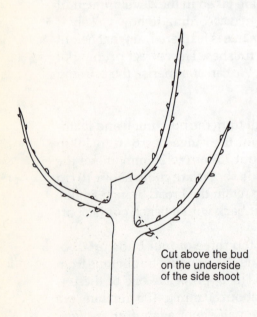

Cut above the bud on the underside of the side shoot

Figure 6.7 Bench pruning for training central leader trees

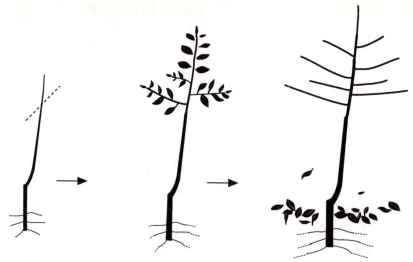

Figure 6.8 *The development of a two year old cut tree*

become too bushy and shaded cut back to a suitable new leader and remove the excess growth with one cut. The lowest side shoots on the tree should be maintained at no lower than about 600 mm from the ground.

(b) Some nursery trees will have too few side shoots or have them in the wrong place. All unsuitable side shoots should be cut back with bench cuts. The leader is then headed back to an appropriate height which is not

Figure 6.9 *Central leader peach trees on peach seedling rootstock. Height is about 3m*

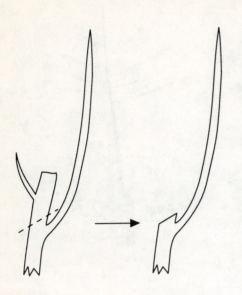

influenced by by unsuitable side shoots. Several "leaders" will develop as before and should be treated in the same way. That is, select the leader, and pinch or cut back the others. By mid-summer you should be able to choose the developing side shoots which are suitable and cut the rest off. From then the pruning described for the previous example applies.

(c) If the nursery tree is not feathered at all, that is, there are no side shoots, all that is needed is to head it back to about 600 mm above ground level. Several leaders will develop. When they are about 80 mm to 100 mm choose the leader and cut or pinch back the others.

Figure 6.10 *Because of the nature of apical dominance a chosen leader must be the strongest available shoot and at the top of the limb*

(d) If the nursery tree is very short (e.g. about 600 mm) it should be headed back to about 400 mm. Early in the spring the leader is selected and the others pinched completely back as no side shoots are wanted this close to the

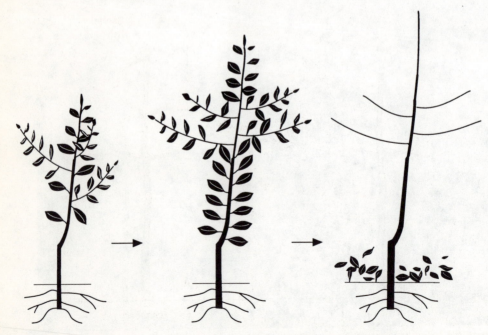

Figure 6.11 *The development of a one year old apple nursery tree*

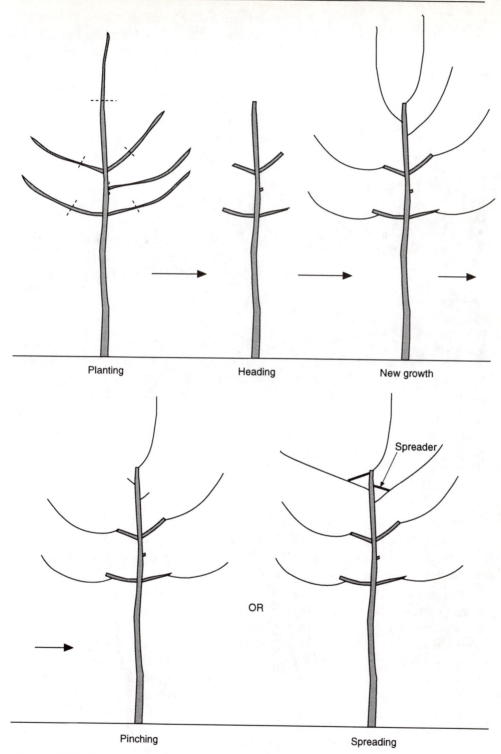

Planting   Heading   New growth

Pinching   OR   Spreading

Spreader

Figure 6.12 *Training a feathered nursery tree as a central leader*

Figure 6.13 *Using poles to support the leader in apple trees on very dwarfing rootstock. Laimburg Research Station, South Tyrol, Northern Italy*

ground. In most cases the strengthened leader will feather as it grows. The best are kept and the others shoot at a broad angle. Apical dominance is now controlling the tree.

In trees that do not develop side shoots well, scoring above the appropriate bud with a budding knife is usually enough. This temporarily stops the downward flow of the auxin and kick starts the selected bud. If not, another option is the use of the plant growth regulator Cytolin which contains a gibberellin and a cytokinin. If this is painted onto buds in the spring according to the directions even the most difficult trees can be manipulated.

It is critical with all of the above procedures to:
- Head back to compensate for transplant shock and to put the root and shoot systems in balance.
- Select the most suitable leader which results and remove its competition by cutting or pinching back. This effectively prevents them from

ever competing again as leader, as dominance has been established, but they can grow back as side shoots which develop into fruiting arms.

- Select the best of the side shoots and remove the rest. The earlier this is done the quicker you have control over the tree's growth.

The procedures described above apply to most fruit varieties on most stocks. However trees planted on dwarfing stocks commonly need additional management. This is because the strength of the dominance is directly related to the strength of the stock. Another problem is the poor anchorage of dwarfing rootstocks.

So why use dwarfing stocks? There are several reasons:

- Early bearing. It is possible to achieve production in year two or three. (Economic production can only be achieved with many trees per hectare and very close planting distances.)
- Reduced production costs. If all mangement procedures on the tree can be achieved from the ground, production costs are reduced. As a rule of thumb for every metre we climb a ladder to manage the tree, production costs are doubled.
- Better sunlight penetration. This leads to consistent and high fruit quality.

There are prices to be paid for this earlier and cheaper production:

- very high capital costs
- poorer tree anchorage commonly needing support
- much less margin for error
- more precise training procedures requiring a far better understanding of tree growth
- much better water and nutrient management
- very effective weed control.

A large number of variations of the central leader style have been developed. Often the difference between the systems is a matter of detail or of the dominant rootstock used. These described below are among the better known.

Spindle bush

The spindle bush was a simple central leader tree on semi-dwarfing rootstocks planted in between "real" vase apple trees to increase early orchard production. Most were freestanding. Planting density was 1000 to 1500 trees/ha.

Slender spindle

The slender spindle was a more intensive system for apples developed in Holland in the 1950s to replace the spindle bush. Slightly more dwarfing rootstocks were used to allow for higher planting densities from 1500 to

Figure 6.14 *Slender spindle apple trees each with their own support posts and support ties to stop fruiting arms hanging below the horizontal under load*

3000 trees/ha. Basically the trees were a little smaller and more open than the spindle bush system.

There was much variation because of experimentation with stocks (but mostly M9), interstems, tree support, budding height and nursery trees specifically grown. A significant advance was the training of the side shoots to specific angles.

The training consisted of the elements discussed previously (heading, choosing a leader, removing the competition). Tree size was controlled by cutting back to a lower leader each winter. All side branches were trained to near horizontal positions and no upright growth (apart from the leader) was tolerated. When the side shoots became too strong or too long they were cut back to a stub and regrown. Other shoots may be cut back to a new sideways leader. In any one year there was little pruning but much training. The system produced small trees which were in full production by year three and from that time the crop controlled the tree size.

North Holland spindle

The North Holland spindle tree is narrower, more heavily pruned and shorter. The leader is often bent to reduce vegetative growth and produce smaller trees. This system is often planted in multi-row beds. Even more dwarfing rootstocks than those required for the slender spindle can be used.

Tyrol slender spindle

For the higher light intensities and warmer temperatures of the south Tyrol (in northern Italy) the system is modified. The leader is cut in half each winter after the tree has reached its required height, to encourage branching. Branches are tied horizontal or lower to reduce tree vigour. At intervals the tree height is reduced by cutting back into older wood and selecting a weak leader. For spur type varieties the leader is headed every year and competing narrow angle shoots are removed. Spreading and tying is usually necessary.

Vertical axis

The vertical axis (also called the axe system) was developed in France. Trees are on dwarfing rootstocks and are planted very high density (1500/ha - 2000/ha) in single rows. This system encourages the tree to grow its own way to quickly achieve a balance between fruiting and vegetative growth.

Apical dominance is weak because of the dwarfing rootstock, and the unheaded leader must be supported by tying to a trellis, to a suspended bamboo pole or a post. Pruning is minimal and the tree reaches its maximum size in two to three years. The trees are much higher than the spindle trees.

The trees are **not** headed when planted to gain maximum benefit from apical dominance but careful tree management is necessary to avoid transplant shock. Pruning consists largely of preventing lower shoots competing with the leader (especially with spur bearers) and to keep maximum sunlight penetration into the lower canopy. Two related French systems (l'axe differe and the fuseaux SX) reduce the tree height when the framework has been established, allowing for all management to be performed from the ground.

Figure 6.15 *Vertical axis apple trees. Note the bamboo support poles hanging from a trellis wire to support the weak leader. The stock is M9*

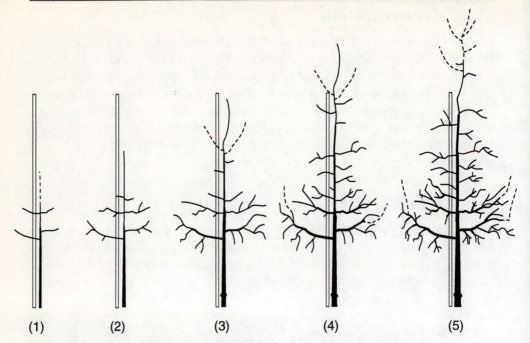

Figure 6.16 *Pruning and training of vertical axis trees planted with branches and supported with a 3 m pole.*
(1) At planting, prune about 300 mm above the top usable branch and tie the trunk to the support pole.
(2) During the first year tie the unpruned leader vertically.
(3) In the late spring of the second year remove vigorous shoots (dotted) and during the summer, tie the unpruned leader vertically.
(4) In the third year continue pruning and training as in the second year; it may be necessary to shorten lower scaffolds by pruning into older wood.
(5) In the fourth year continue pruning and training as in the third year

USA central leader

The USA central leader is a common training system in the north-west of the USA, in the very hot dry areas of Washington and Oregon states. Trees are on semi-dwarfing stocks and much bushier to protect against the very high light intensities and heat. This system is a simple central leader tree but much bigger than the European trees.

HYTEC

In Washington state another closely related system, the HYTEC, has been developed to suit the local conditions. The system is very much like the vertical axis but attempts to produce less shading in the tree. The most common rootstock is M9 (or M7 with weaker growing varieties such as Braeburn or Spur Red Delicious).

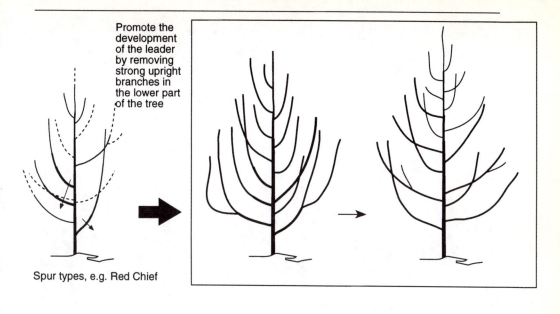

Promote the development of the leader by removing strong upright branches in the lower part of the tree

Spur types, e.g. Red Chief

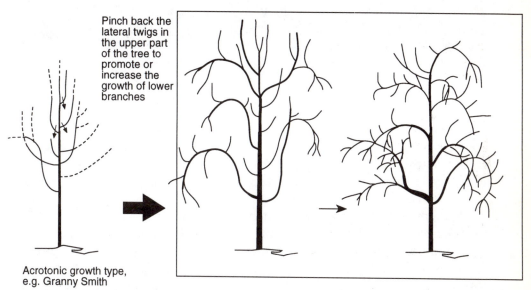

Pinch back the lateral twigs in the upper part of the tree to promote or increase the growth of lower branches

Acrotonic growth type, e.g. Granny Smith

Figure 6.17 *Channelling the trees growth habit into the "central axis" form*

As in most other central leader systems the trees are headed when planted and excess side shoots are removed. The leader is selected and competitors are removed. The leader is tied to a support which is usually a suspended bamboo pole. In later years pruning back of the leader to a lower leader is used. Bending the leader is also used to reduce dominance once the framework is established. Side shoots are tied down to encourage cropping and discourage competition with the now weakened leader.

Figure 6.18 *Central leader apple tree (Red Chief Red Delicious) about two metres high*

Multi-row beds and other developments

All of the above systems are variations on the central leader training system. There are two main areas of difference between the types. One is the rootstock used, which in large part controls the size of the tree. The other is the degree of training and tying side branches to horizontal to induce early production and also to help control tree size.

Refinement of this training system has led to some innovative approaches to orchard design. One of these is the multi-row bed design.

In an attempt to achieve very early economic levels of apple production of consistently high quality fruit on very expensive orchard land, the multi-row bed system was developed in Holland in the 1980s. The trees were propagated on very dwarfing rootstocks such as M9 and M27 and trained to the north holland spindle. Because the anchorages of these stocks are very weak each tree is supported by a post. Side shoots are tied into place to induce early cropping and to minimise water shoot development. Fruiting arms are tied up so that they do not drop below horizontal with the fruit load.

While early orchards were planted as single rows, double row beds of 2000 trees/ha and three row beds with 2400 trees/ha are grown.

As confidence in the multi-row bed systems was gained tree densities were increased so that two row beds with 3000 trees/ha were developed.

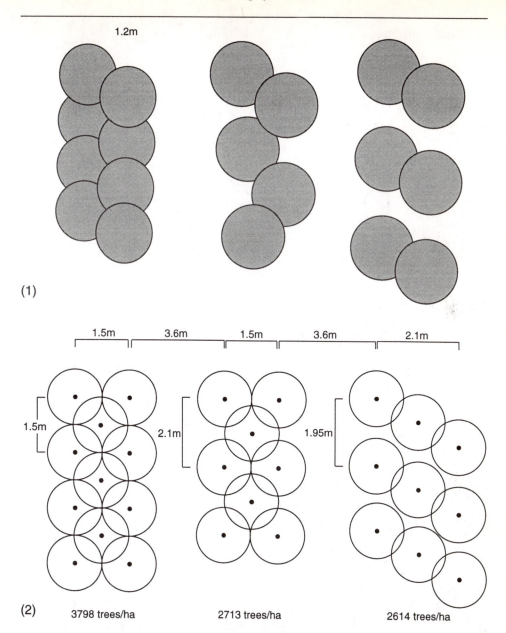

Figure 6.19 *(1) Examples of double-row tree arrangements. The double-row design on the left has poor sunlight distribution, worker access and penetration of sprays. The middle and right designs have improved access, but have lower tree densities. The design on the right has cross paths for improved worker access and spray distribution (the figure is to scale with 1.5 m diameter trees)*

(2) Examples of trees arranged in three-row designs. The design on the left has very high tree density, but extremely poor access to the centre of the bed. The middle design is an improvement, but worker access to the centre of the bed is still poor. The design on the right, with cross paths, has excellent worker access, improved spray penetration and better light distribution. To achieve proper worker access, tree density is reduced (the figure is to scale with 1.5 m diameter trees)

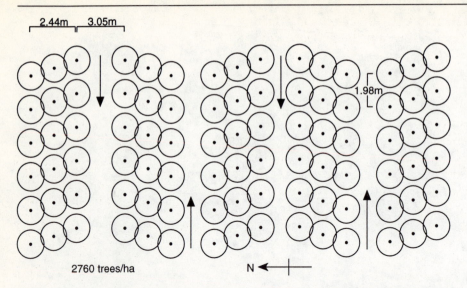

Figure 6.20 *Three-row bed system with cross paths. Arrows indicate sprayer direction. A tractor alley is oriented east/west, and cross rows in the bed are oriented approximately north/south for uniform light distribution.*

A number of three row bed planting patterns were developed with densities from 2300trees/ha to 4000 trees/ha.

Various planting designs were tested not only for production but also to test spray and light penetration. In addition specialised weed and tree spray equipment was designed and mechanically assisted harvest aids were introduced.

Four row beds and five row beds were developed with 3000 trees/ha to 4000 trees/ha. Some of these designs were heavily promoted by the nursery industry in Belgium and Holland. There was some evidence that weed and tree spray penetration was less than adequate. But the major question

Figure 6.21 *Five row beds of Jonagold on M9 in Belgium.*

mark about these systems is the intense shading which develops from about year five (depending on the specific planting pattern). It is highly unlikely that multi-row bed systems will prove appropriate for Australian conditions. Nursery tree costs are much higher in Australia as well.

Other developments of interest from Europe have come in recent times. One is the discovery of a MacIntosh apple cultivar genetically unable to branch. The cultivar called the Wijcik was developed at the East Malling Research Station in England. Such trees can be planted very close together with no development of shading as the trees grow. These have been marketed in Australia as "Ballerina". If the gene can be transferred to a more commercial variety there may be some potential.

Figure 6.22 *Five row beds of Jonagold on M9 in Belgium. Note the training method.*

Another is the development of extremely high density orchards. At the East Malling Research Station, a meadow orchard was trialled where the trees were grown for one year at planting distances of 300 X 300 mm. The next year the trees cropped and the tree and crop were harvested like a wheat crop.

A super high density apple orchard was planted at Geldermalsen south of Amsterdam with a block of trees planted 0.3m by 1.0m in multi-row beds with traffic ways at intervals through the block. The planting density was 40,000 trees/ha. The trees were irrigated, fertilised and protected from weather in a computer controlled all weather hall where the plastic roof opened and closed as the computer detected clouds. Total capital cost is over one million dollars per hectare.

Figure 6.23 *Wijcik apple tree in Holland. This cultivar lacks the gene for branching*

Figure 6.24A *Super high density (40,000 trees/ha) of Jonagold on M27 stock in Holland.*

Figure 6.24B *Row spacing and in-row tree spacing 18 months after planting*

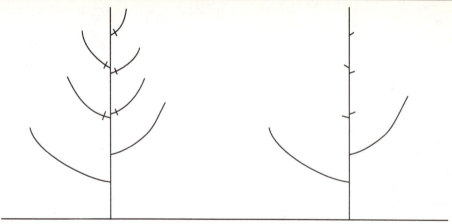

Figure 6. 25A *Development of a one leader palmette*

Palmette

The palmette tree is trained to a two dimensional fan shape with one or two (sometimes three) leaders per tree. This system is very popular for stone

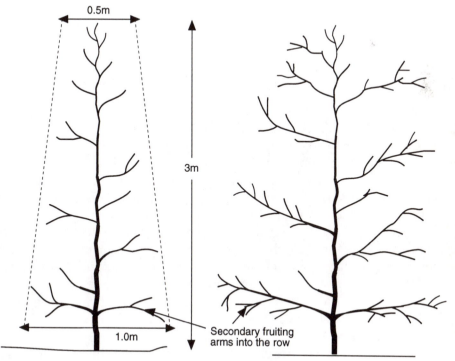

Figure 6.25B *End and side views of a mature palmette canopy showing lateral placement*

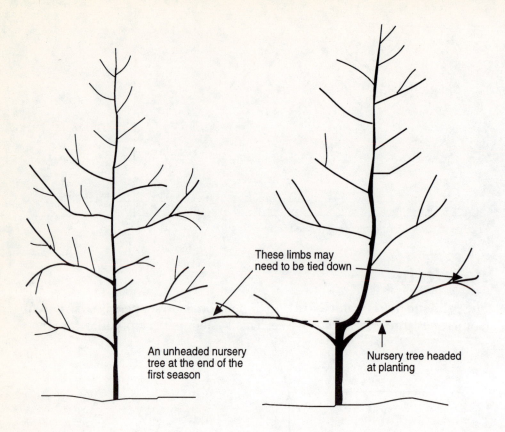

These limbs may
need to be tied down

An unheaded nursery
tree at the end of the
first season

Nursery tree headed
at planting

Figure 6.26 *Developing the branch framework of a palmette tree during the first season*

fruit especially peaches and nectarines. Because these fruits develop
"feathered" shoots which cause shading, the narrow thickness of trees in
this system assists light penetration.

The palmette tree is about 2.5 m to 4.0 m in width along the row and
about 0.5 m each side of the tree line (that is 1 m thick). The height is selected and controlled by pruning.

As peach trees usually arrive from the nursery well feathered, pruning
at planting is very simple. If the tree is a single feather then use it as a single
leader palmette simply by **not** heading it but removing most side shoots
except those which have been chosen to develop as fruiting arms.

If a two leader palmette is required then cut back to two appropriate
leaders and remove all competition by cutting back to a basal bud. Do not
head the leaders.

If the nursery tree has three feathered shoots remove the one which is in
the least useful position and cut back most side shoots on the other leaders.
Do not head the leaders.

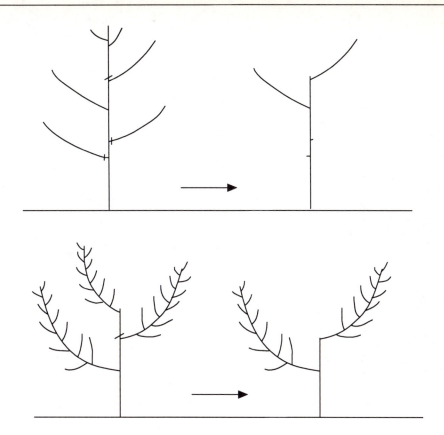

Figure 6.27 *Development of a two limb palmette*

In the second and subsequent years all that is needed is to maintain the dominance of the leaders, to cut back short excess side shoots and to develop fruiting arms at intervals of about 0.7 m apart, most of which should be directed into the traffic way. A suitable framework will be established by about year four. If the tree is too high at this time then head it back to a lower suitable leader. By leaving the leader unheaded until this time maximum framework development will have occurred.

Most trees do not need support such as a trellis if developed as a palmette. Trellising may be needed where anchorage is poor, wind loads are high or very specific training is needed.

There are many variations on the basic palmette style. For example it has been common in the past in Europe to use **oblique palmettes** with fruit varieties which do not have dwarfing rootstocks e.g. pears. The main leader and branches are trained to low angles, greatly reducing apical dominance. In the French Marchardt system, pear trees are planted at 45 degrees to the vertical along a vertical trellis. This method is also called the Belgian cordon.

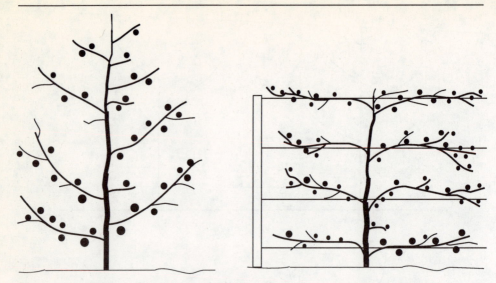

Figure 6.28 *(a) Fruit distribution of a freestanding palmette peach tree. (b)Fruit distrubution of a trellised palmette peach tree*

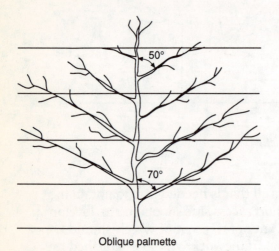

Oblique palmette

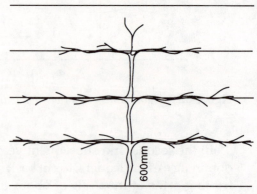

Horizontal palmette

Figure 6.29 *Training for hedge shapes*

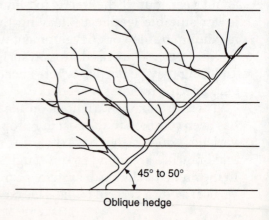

Oblique hedge

Figure 6.30 *French Marchardt palmette system with pear trees controlling the size of very vigorous trees*

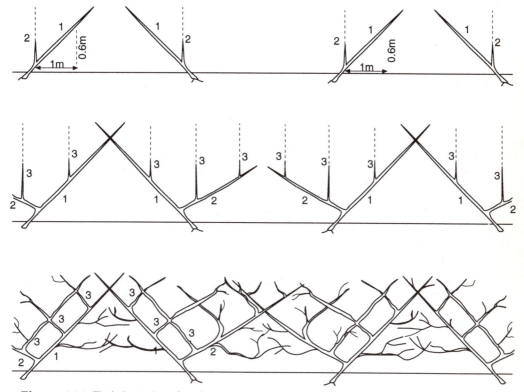

Figure 6.31 *Training a Bouche-Thomas hedge*

A variation on this theme is the **Bouche-Thomas hedge** where the tree is planted at a 45 degree angle but is trained in both directions.

Another variation on palmette is the **espalier system** often used in gardens, garden orchards and other situations where space is limited and a support is available. The system is to simply fasten branches at intervals onto the support which could be the wall of a house, a fence, a pergola or any other structure.

Trellis type training systems

These training systems require trellis and wire support not simply because of poor anchorage or need to support a fruit crop. The trellising is needed because the tree is to be trained into a shape which it would not normally adopt. The tree is to be tied to the trellis for the purpose of imposing a shape upon it. A common reason for doing this is to obtain early cropping.

Tatura trellis

This system was developed at the Tatura Research Station near Shepparton in Victoria for the purpose of assisting in the mechanical harvesting of processing fruit especially canning peaches. A secondary aim was to allow for mechanical pruning using a hedging machine. Taturatrellis has subsequently been successfully used for a wide range of fruit crops such as apples, cherries, plums and nashi.

The major drawbacks of the system are:

* high capital cost ($30,000/ha)
* labour intensive training requirements.

Despite this the mini-Tatura is one of the best ways of training nashi which have very brittle wood and so need support when cropped. The Tatura trellis has been very successfully used to obtain early and high production from cherry trees propagated on vigorous rootstocks.

However, fruit size drops as the canopy is filled and annual remedial pruning is necessary to minimise this problem.

The trellis consists of either:

* Wooden posts (commonly DNOC treated pine) 100 to 125 mm in diameter and 3.6 m long sunk into the ground at 60 degrees and bolted where they cross.
* Agbar or similar frames on posts and on horizontal plates.

The distance between the support assemblies is usually 10 m. Tree spacing is 1.0 m to 2.0 m along the row with 4.5 to 6 m row spacings giving tree densities from 1500 to 2500/ha. Very substantial anchor assemblages must be constructed at each end of the rows to take the strain of the tensioned

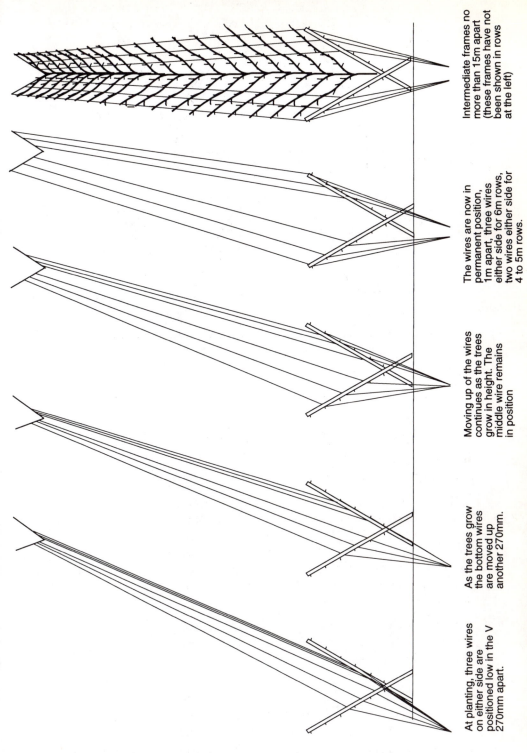

At planting, three wires on either side are positioned low in the V 270mm apart.

As the trees grow the bottom wires are moved up another 270mm.

Moving up of the wires continues as the trees grow in height. The middle wire remains in position

The wires are now in permanent position, 1m apart, three wires either side for 6m rows, two wires either side for 4 to 5m rows.

Intermediate frames no more than 15m apart (these frames have not been shown in rows at the left)

Figure 6.32 *Positions of wires on the Tatura trellis as the trees grow*

113

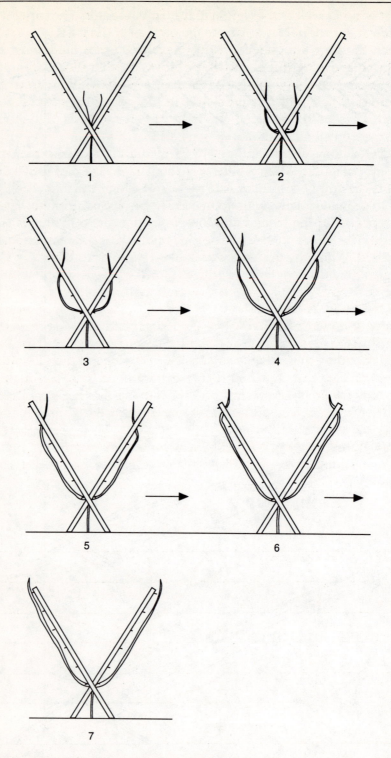

Figure 6.33 *Leader training on the Tatura trellis*

support wires as well as the weight of the fruit in season. The support wires are fitted at appropriate intervals to the outside of the posts.

When the trees are planted, if they have two limbs these must be oriented in a plane at right angles to the tree row. All other shoots are removed. If the nursery tree is a whip then it is headed at a point level with where the posts cross. When growth commences in the spring only two shoots are retained. These will be trained into branches on each side of the trellis. In the first season all other shoots are removed.

The two shoots are encouraged to grow initially in the normal apically dominant positions. When the shoot is about 1.0 m to 1.5 m above the first wire it is carefully bent under and to the outside of the first wire. It is allowed to grow normally again to maximise dominance growth. The process is repeated at the second and then in turn at the other wires (see the diagram). The leader must not be headed until the tree framework has been established. When the leader is developing, some side shoots are allowed as long as they do not compete with the leader. These will be required to form the canopy which carries the fruit on each side of the trellis.

It should now be obvious how this training system works by manipulating apical dominance. Growing two shoots only and allowing them to grow vertically in the first season produces the maximum length and thickness of the limb. Bending these shoots outside the first wire and then allowing vertical growth continues maximum growth controlled by apical dominance. By the time the limb has reached past the topwire the framework has been created in the shortest possible time simply by using the tree's natural apical dominance.

While this framework is developing other shoots must not be allowed to compete with the leader. But other shoots are encouraged to develop. The

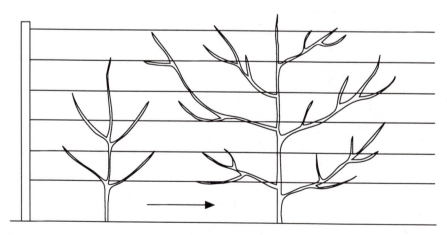

Figure 6.34 *Branch training for the Tatura trellis*

aim is to fill the canopy with fruiting arms at about the same time as the leader has reached the top of the tree. If these shoots are encouraged to grow at angles then the dominance is reduced.

The 60 degrees of angle reduces the apical dominance and hence vegetative growth. This angle is adequate to promote the development of fruit bud. By the time the leaders have reached the top wire (often by year two) the tree will be ready to go into production. The spreading of the side shoots along the wire reduces their dominance even further and these shoots will become fruiting arms very quickly.

Having the wires on the outside of the pos tmeans there is some tendency of the fruit to tear the wire away from the post and spread the tree. But what is far more important is the tendency of the tree to grow vertically for the rest of the season. If the wires were on the inside of the posts, they

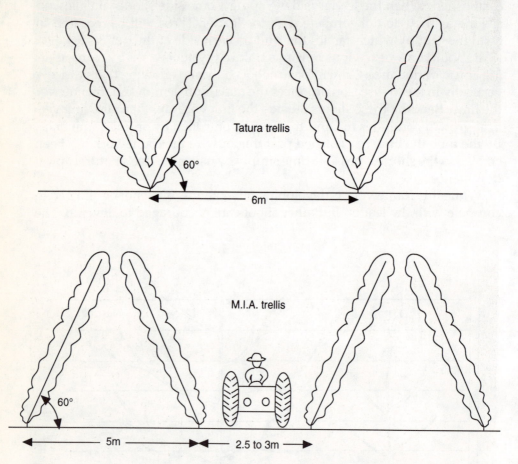

Figure 6.35 *Trellis systems for mechanical pruning and harvesting will let plenty of light in, provided the canopy is not allowed to get too deep by summer pruning and some thinning out of laterals in winter. The Lincoln canopy can have the rows orientated any way. The other two should be planted as near north/south as possible*

116

would be pulled to the centre of the vee by this apical dominance. In practice, especially on cherries, spreaders or stays must be placed between the wires to try and prevent them from being pulled together.

Figure 6.35A *Tatura trellis with plums (Tatura Research Station near Shepparton)*

Figure 6.35B *Tatura trellis with peaches. Mechanised pruning in the USA*

Figure 6.35 *Tatura trellis with cherries. This method of traiming, although somewhat costly, is very effective for training strongly apically dominant cherry trees for early and high production (Tatura Research Station near Shepparton)*

Once the tree commences production the main pruning and training requirements are:

- To keep the centre of the vee of the trellis open by pruning off vertically growng shoots or by bending them outside the wires.
- To spread the side branches along the developing canopy to produce a wall of fruit bearing surface.
- To prune out excess shoots in this canopy to prevent excessive shading.
- To control the crop by pruning off excessive fruit bud especially that on the least vigorous growth. In cherries the best time to do this is in the spring.

M.I.A. trellis

The M.I.A. trellis is very similar to the Tatura trellis, but it is arranged as in Figure 6.36.

The tree rows are planted parallel to each post row and at 60 degrees above horizontal. The manipulation of each tree is the same as in the Tatura trellis except each tree has one leader instead of two.

These trees are essentially palmettes, but leaning at 60 degrees. A major advantage is that spraying and other tree management can occur from both sides of each canopy surface.

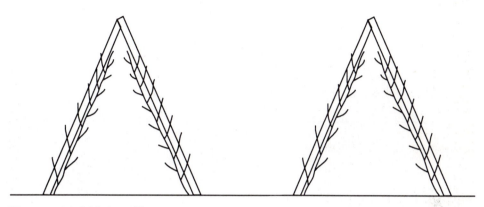

Figure 6.36 *M.I.A. trellis*

Figure 6.37 *M.I.A.trellis with apples at Orange, NSW*

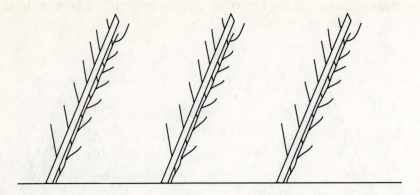

Figure 6.38 *Yanco cantilever*

Yanco/Tatura cantilever

This system is essentially half of either of the previous two. It has found little commercial application.

Canopy Types

Canopy systems are very specialised trellised systems. The trellised systems previously discussed modify apically dominant growth slightly. Canopy systems depend on very modified growth patterns. Four systems are discussed under this heading but there are many other variations. One of these canopies was developed initially to cater for mechanical harvesting and pruning of fresh fruit varieties. The other three are different approaches to the concept of manipulating tree growth for maximum production and/or fruit quality.

The Lincoln canopy

The Lincoln canopy was developed at Lincoln University in New Zealand in an attempt to allow for mechanical harvesting and pruning of fresh fruit. The system uses a tee shaped trellis. Vertical posts are planted at 10 m intervals about 1.2 m into the ground and 1.5 m above ground. The top of the tee is a 3 m rough sawn timber 100 x 75 mm. Strong anchor points planted at the end of each rows or box strain assemblies are built at each end.

The trees are planted 2.0 to 2.4 m apart along the row and the row spacing is 4 m . This gives a density of 1040 to 1250 trees/ha.

The training depends on the manipulation of apical dominance. Nursery trees as whips are essential for this system. Four or five shoots (one as a

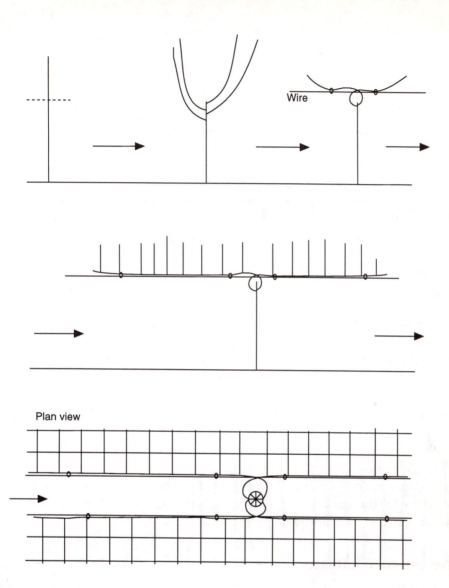

Figure 6.39 *Lincoln canopy*

spare) are encouraged to develop as competing leaders. When these leaders are about one metre long they are bent over and tied to a wire at about half length using tools such as a Max Tapener. Any fifth leader is removed at this stage.

At the end of the season all the leader growth is tied down. In the next season a large number of shoots will develop from buds on the top of the tied down shoot. Its apical dominance no longer exists because it is

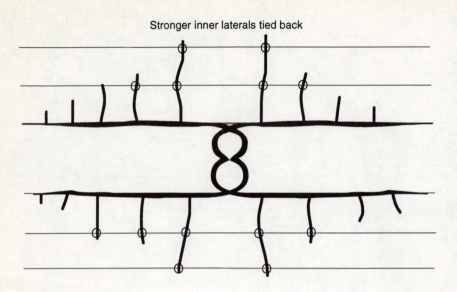

Figure 6.40 *The Lincoln canopy*

horizontal. As the vertical shoots reach a length of about 1.3 m on average, tying down will begin.

The tree is now ready to start bearing as the now horizontal shoots have lost their apical dominance, allowing fruit bud to develop. Vertical shoots

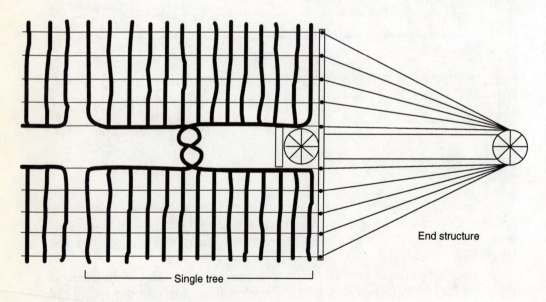

Figure 6.41 *Canopy training system*

Figure 6.42 *Lincoln canopy in winter showing the fruiting arms arrangement*

will again develop during the next season. Some of these will be selected for tying down to fill gaps. The rest can be mechanically pruned off using a hedging machine.

Very high levels of production can be achieved very early in the life of the trees (e.g. 25 tonnes of Red Delicious/ha in year four and up to 60 tonnes

Figure 6.43 *Lincoln canopy at harvest time with apples. Lincoln, New Zealand*

in year six). Poor fruit bearing shoots can be removed at intervals and replaced by fresh young tied down shoots. A big advantage of this system is that all of the fruit is produced in a narrow horizontal layer. If excess shading is avoided by hedging the vertical shoots, all fruit gets even and adequate light resulting in a crop uniform in size, colour and maturity.

The Solen

The Solen is an apple training system developed in France by the innovative researcher Jean-Marie Lespinasse. The system is another approach to the training of high density orchards on dwarfing rootstocks. The stock used is M9 and the trees are planted 2 x 3.2 m in single rows or (4 + 2.4 m) x 2 m in double rows.

Support posts are planted along the row at 6 m intervals with two wires at 1.4 m and 1.5 m above the ground. When the tree is planted as a whip it is headed at 1.1 m above the ground and two new leaders are allowed to develop. These are initially supported vertically on wires then at 45 degrees in late summer and then horizontally in the winter. Apical dominance is used to grow the stem and then its dominance is removed by tying it horizontally.

During the next two years, pruning and training consists of removing all vertical and unsuitable shoots. The only shoots retained are those with broad branch angles to the side of the original shoots. The result is similar to the four armed Lincoln canopy but with two arms and no tee-bar support. Productivity and fruit quality to date are similar to the those of the vertical axis system.

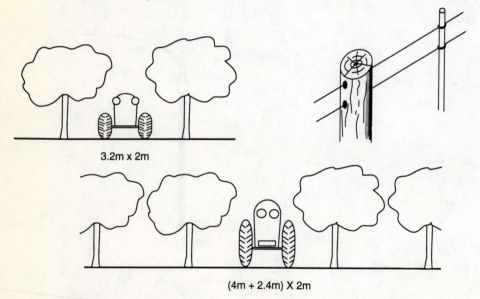

3.2m x 2m

(4m + 2.4m) X 2m

Figure 6.44 *Solen training*

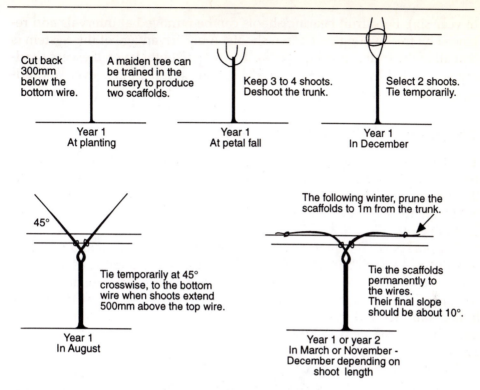

Cut back 300mm below the bottom wire.

A maiden tree can be trained in the nursery to produce two scaffolds.

Year 1
At planting

Keep 3 to 4 shoots. Deshoot the trunk.

Year 1
At petal fall

Select 2 shoots. Tie temporarily.

Year 1
In December

45°

Tie temporarily at 45° crosswise, to the bottom wire when shoots extend 500mm above the top wire.

Year 1
In August

The following winter, prune the scaffolds to 1m from the trunk.

Tie the scaffolds permanently to the wires. Their final slope should be about 10°.

Year 1 or year 2
In March or November - December depending on shoot length

Figure 6.45 *Solen training — bending of the two scaffolds*

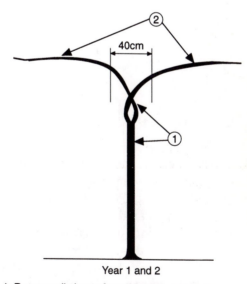

②

40cm

①

Year 1 and 2

1. Remove all shoots from the trunk and arches.
2. Remove or cut back, if necessary, the most upright shoots on the scaffolds to favour the development of those with an open crotch angle.

Figure 6.46 *Deshooting*

These operations both should be done at petal fall.

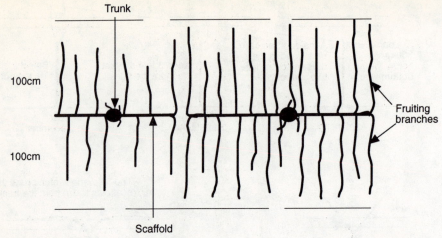

Figure 6.47 *Deshooting viewed from above*

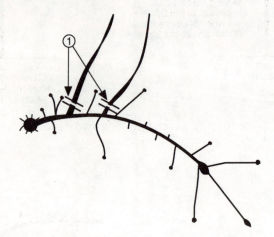

Figure 6.48 *Pruning*

The Ebro-espalier

The Ebro-espalier is a patented training system developed in New Zealand by Roger Evans. The system is a little like a narrow four layered Lincoln canopy. The trellising system is a complex set of four layers of wire (with six runs of wire in each) with base plate trellis support frames at 25 m intervals. The end frames are larger and heavier to support the tension strain of the wires as they change direction down to the end assembly (the anchor).

The trees are planted 2.5 m apart along the row and the row spacing is 3.6 m which gives a planting density of approximately 500 trees/ha.

The trees are headed at planting and four leaders are selected, two at each side of the trellis system. The vertical shoot from each leader must be branched into three at each level of the support wires in the trellis. One is trained to continue vertically and the other two are bent over in opposite

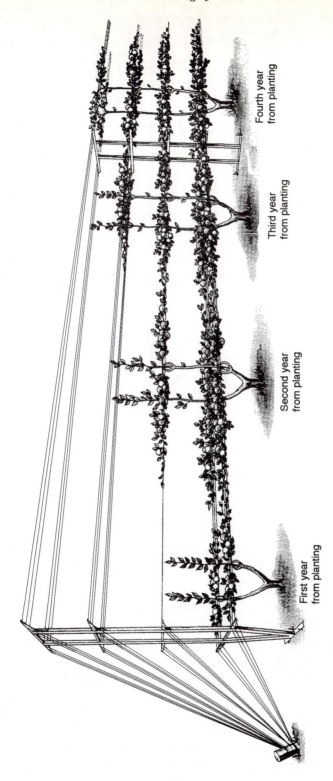

Figure 6.49 *Development of trees on the Ebro-espalier*

127

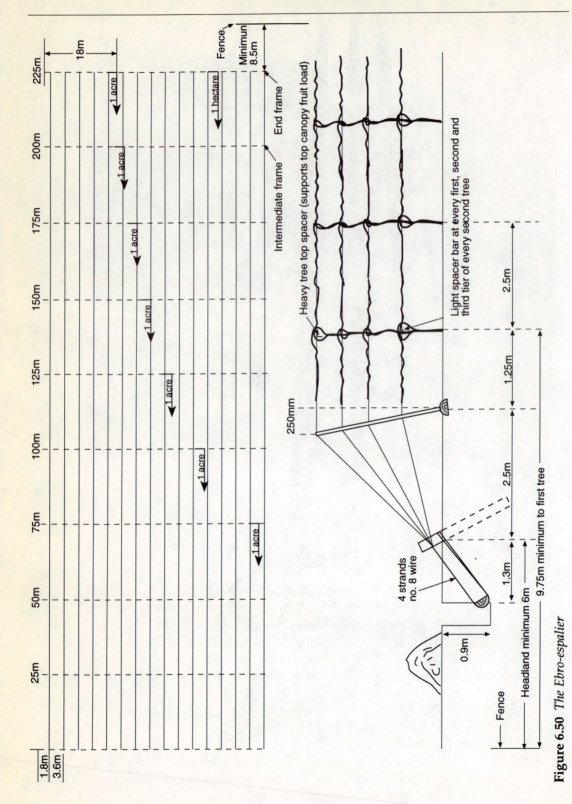

Figure 6.50 *The Ebro-espalier*

Figure 6.51 *Ebro-espalier. New Zealand*

directions along the trellis support wires and tied down. This is repeated at each level in the trellis system.

There is no doubt that the system is very productive. However there are five major problems which prevent it from being more widely accepted:

(1) Apical dominance will produce multiple vertical shoots from each horizontal branch at each level. These shoots produce massive shading which cannot be mechanically pruned because the wires and the trellis supports are in the way. Multiple detailed pruning is necessary.

(2) Unless this shading is controlled, fruit size, colour and quality are reduced. Because of the detailed hand pruning, the cost of production is high.

(3) Because of the design of the trellis much of the work on the system must be done on ladders or height supports.

(4) The labour cost of training is very high.

(5) The capital cost (including patent costs) is the highest of all the trellis systems.

Figure 6.52 *Japanese canopy with nashi. Note the planting density.*

The Japanese canopy

Because of the domestic demand for large fruit of very high quality (high visual quality, at least) in Japan, a number of tree management systems have been developed. One of the most widespread is the Japanese canopy.

The trees are not planted on dwarfing rootstocks, so tree density is low at about 100-200 trees/ha. Early production and high production levels at tree maturity are not considered priorities as these aims are seen to be achieved at the expense of fruit quality.

The trees are headed at about 1.5 m and the new leaders which develop are tied down horizontal towards the end of the growing season before they become too woody. A coarse wire net is spread over the orchard and is used to tie the trees down. So the system has some similarities to the Lincoln canopy; that is, leaders are grown vertically using apical dominance and then tied down for production. In the Japanese system the tree branches are tied down radially to the main stem.

The canopy of each tree covers an areas up to about 10m radius so the trees are very big by modern standards. The entire orchard area is covered by the trees' fruiting canopy. All of the fruit is grown in one plane and if the trees are well managed all receive the same amount of light.

What makes the system unique is the crop management. The crop is very heavily thinned so that productivity is low by world standards. The aim is to grow the largest fruit possible. The fruits are grown in special paper bags for most of the season usually with three phases of bagging being used. The main reason for the bagging is to achieve perfection in the skin of the fruit. This includes the quality of skin texture and bloom. It also includes the unique development of colour. When fruit is grown in a bag it does not develop the green colour but becomes etiolated (pale). When the bag is removed for the maturation process the skin develops unique colour in terms of shade and intensity. This feature is demanded by the Japanese consumer.

This technique is used on apples, nashi, table grapes, persimmons and other fruits. While it is labour intensive and expensive, the financial returns justify it in Japan. The technique also ensures that there is no pest or disease or weather mark (e.g. russet) on the fruit and no pesticide residue either.

There is a disadvantage. As discussed earlier, fruit is a major photosynthesising organ which contributes to fruit size and quality

Figure 6.53 *Japanese canopy with table grapes. Each bunch returns the grower between A$25 and A$50*

parameters. For size the Japanese thin their crops very heavily. But there is no substitute for the other factors. The bagged fruit is much lower in sugars than unbagged fruit (at least 2 degrees Brix less) and is much lower in flavours and perfumes. Japanese fruit in general is bland and flavourless.

CHAPTER 7

Rootstocks

Commercial fruit trees are two trees in one. One variety called the rootstock or stock produces the root system. The other variety called the scion is budded or grafted onto the stock and produces the fruiting part of the tree. In modern orchards it is critical that the appropriate combination of stock and scion is selected for the environment and orchard design.

The stock is selected because of its contribution to the tree's development in the following areas:

- tree size
- tree shape
- precocity
- compatibility with the scion
- suitability to the local soils
- suitability to the local climate
- pest and disease resistance
- tree anchorage.

For some fruits (e.g. apples) substantial research has produced a very large number of stocks to suit specific requirements for any combination of growing conditions. The advent of plant variety rights legislation in Australia (as well as non-propagation agreements) has meant that a range of these stocks is becoming available. Some still have to be trialled under local conditions to determine if overseas results also apply to Australian conditions. There is also some question as to whether the stocks and scion varieties imported into Australia are genuinely those which we see overseas.

For other fruits the range of stocks available in Australia is more limited.

The following is a brief introduction to the major stocks available in Australia. Some which are not yet available, but which are likely to be available in the near future, are also included.

Apples

The following list of major apple stocks commences with the most dwarfing:

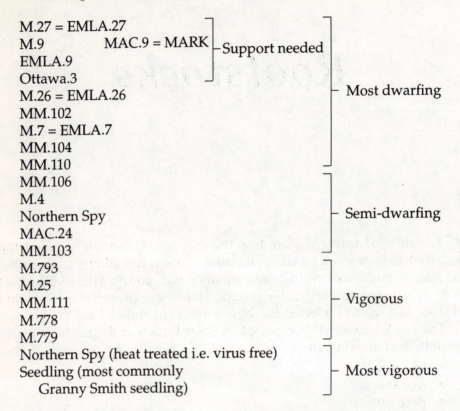

M.27 = EMLA.27		⎤
M.9 MAC.9 = MARK	— Support needed	
EMLA.9		
Ottawa.3	⎦	— Most dwarfing
M.26 = EMLA.26		
MM.102		
M.7 = EMLA.7		
MM.104		
MM.110		
MM.106		
M.4		
Northern Spy	— Semi-dwarfing	
MAC.24		
MM.103		
M.793		
M.25		
MM.111	— Vigorous	
M.778		
M.779		
Northern Spy (heat treated i.e. virus free)		
Seedling (most commonly	— Most vigorous	
Granny Smith seedling)		

In addition to the above list the P series (from Poland), the Budagovsky series (from Russia), the Ottawa series (from Canada), the Bemali series (from Sweden), Jork (from Germany) and the Aotea series (from New Zealand) are being trialled overseas.

Arguably the most important of these at the present time would be M.9 or Mark for very high density supported orchards followed by MM.106 for medium sized non-supported trees. However it is important to match the stock to the scion variety for each particular growing environment and the orchard design which has been decided.

Table 7.1 *Tree size and root production of 22 dwarfing apple rootstocks in relation to M.9 EMLA at the end of the sixth year at Wenatchee, Washington, USA*

Rootstock	Trunk cross-sectional area (% of M.9 EMLA)	Cumulative yield/tree (% of M.9 EMLA)		Cumulative yield efficiency (yield/TCSA, % of M.9 EMLA)	
		Golden Delicious/ Granny Smith	Red Chief Delicious	Golden Delicious/ Granny Smith	Red Chief Delicious
M.27 EMLA	41	31	25	80	65
P.16	49	61	44	120	82
V.5-3	63	69	66	127	88
P.22	64	69	52	113	92
Bud.146	67	54	65	96	85
MAC.9	67	82	65	120	91
M.9	72	75	52	102	78
Mark	73	76	44	110	66
P.2	88	77	93	95	109
CG.10	88	77	72	86	92
M.9 EMLA	100	100	100	100	100
V.5-1	109	97	107	105	89
Bud.9	115	116	145	108	113
MAC.39	130	83	133	79	78
O.3	137	100	131	70	100
M.26	149	105	159	81	89
V.5-7	156	114	166	70	120
C6	160	89	104	50	95
V.5-2	181	117	156	66	90
OAR.1	224	51	101	25	43
P.1	240	98	174	44	67
V.5-4	248	124	198	51	77

Table 7.2 *Rootstock vigour. The relevant vigour of apple rootstocks as compared to apple seedlings (100%)*

Pears (*Pyrus communis* "European"; *Pyrus pyrifolia* "Asian" or "Nashi")

At the moment there are only two pear rootstocks in common use in Australia.

Pyrus calleryana (D6) is the rootstock in most established pear orchards in Australia. It is tolerant of a wide range of growing conditions. The major problem with D6 is its vigour. In good soils it grows very large trees which results in pruning and training problems.

Cydonia oblonga or quince has been used in Australia in an attempt to control tree size. To date results have not been good with poor tree growth, low yields and incompatibility problems. Clones being trialled in Europe may provide better size controlling stocks.

Peaches/Nectarines (*prunus persica*)

Prunus persica (peach) stocks

Elberta seedling has been the standard rootstock for many years. It needs a well drained soil. Golden Queen is also widely used.

Rubira (from INRA France) is a red leafed peach.

Monclar (from INRA France) is a vigorous clone.

Hybrids

Nemaguard is resistant to root knot nematode.

60EB160 is a peach/almond hybrid suitable for poor soils.

GF 677 is a peach/almond hybrid probably most suited to replant situations.

Citation is a peach/plum hybrid producing 60% size reduction with peaches and showing some promise for size reduction in plums and apricots.

GF 557 is a peach/almond hybrid showing promise on alkaline soils.

Plums (*Prunus domestica* "European"; *Prunus salacina* "Japanese")

Myrobalan (*Prunus cerasifera*) is a drought tolerant stock most suited for light soils. It delays fruit maturity and produces a large size tree.

Myrobalan 29C is tolerant of heavy wet soils.

Table 7.3 *Gembloux series. Dwarfing cherry rootstocks*

Marianna is a plum cross which produces medium sized trees and heavy cropping.

Marianna 2624 is tolerant of waterlogged soils.

Marianna GF 8-1 is tolerant of a range of soil types and is precocious.

Pixy produces trees about 40% of Myrobalan size and is very precocious.

Apricot *(Prunus armeniaca)*

The apricot seedling produces a large tree and is widely used as an apricot stock.

Cherries *(Prunus avium)*

Mahaleb (*Prunus mahaleb*) is a common rootstock in areas with light, well drained soil. A major problem is its incompatibility with some commercial cherry varieties (e.g. Van).

Mazzard F12/1 (*Prunus avium clone*) is more common in areas with heavy soil. It produces a large tree which is often slow to come into bearing. It is somewhat resistant to bacterial canker and is compatible with all sweet and sour cherry cultivars.

Colt (*Prunus avium* X *Prunus pseudocerasus*) is a little less vigorous than F12/1 and is also somewhat resistant to bacterial canker. It induces broad branch angles in the tree and crops earlier than F12/1.

Charger (*Prunus avium* has not been fully tested in Australia but early trials suggest good resistance to bacterial canker and earlier cropping than F12/1.

Not yet tested in Australia are three series of dwarfing cherry rootstocks from Europe.

The first is the Gembloux Series (GM) from Belgium. Experience from the USA suggests that fruit size on these stocks is very sensitive to crop load.

The second is the Giessen Series (GI) from Germany now being marketed in the USA. as the "Gisela" cherry rootstocks. Experience with three stocks from this range suggests that they are not very dwarfing but are very precocious. Fruit size is not as dependent on crop load as with the Gembloux stocks or Mahaleb. There are twelve stocks in this range and the others are yet to be trialled.

The third is the Weiroot Series from Germany which supposedly range from 30% to 60% of F12/1.

Table 7.4 *Relative tree size of Bing cherry trees on various rootstocks in two NC-140 plantings*

Rootstock	Relative tree size by TCSA compared to Mazzard (%)
Colt	161
Gl.148/1	99
Gl.195/1	102
Gl.196/4	117
GM.61/1	47
GM.79	75
GM.9	50
Mazzard	100
M × M.2	116
M × M 46	106
M × M 60	132

Index

Some other titles in the
PRACTICAL FARMING SERIES
Published by Inkata